# Solutions and Tests For The Human Body: Fearfully and Wonderfully Made!

Dr. Jay L. Wile

© 2001, Apologia Educational Ministries, Inc.
All rights reserved.

Manufactured in the United States of America
Eighth Printing April 2011

## Published By
## Apologia Educational Ministries, Inc.
Anderson, IN

## *Printed by*
## Courier
Stoughton, MA

# The Human Body: Fearfully and Wonderfully Made!
Solutions and Tests

## TABLE OF CONTENTS

Teacher's Notes .................................................................................................. 1

### Solutions to the Study Guide

Solutions to the Study Guide for Module #1 ........................................................... 7
Solutions to the Study Guide for Module #2 ........................................................... 11
Solutions to the Study Guide for Module #3 ........................................................... 14
Solutions to the Study Guide for Module #4 ........................................................... 17
Solutions to the Study Guide for Module #5 ........................................................... 20
Solutions to the Study Guide for Module #6 ........................................................... 23
Solutions to the Study Guide for Module #7 ........................................................... 25
Solutions to the Study Guide for Module #8 ........................................................... 28
Solutions to the Study Guide for Module #9 ........................................................... 31
Solutions to the Study Guide for Module #10 ......................................................... 35
Solutions to the Study Guide for Module #11 ......................................................... 38
Solutions to the Study Guide for Module #12 ......................................................... 42
Solutions to the Study Guide for Module #13 ......................................................... 45
Solutions to the Study Guide for Module #14 ......................................................... 49
Solutions to the Study Guide for Module #15 ......................................................... 52
Solutions to the Study Guide for Module #16 ......................................................... 55

### Tests

Test for Module #1 ................................................................................................ 61
Test for Module #2 ................................................................................................ 63
Test for Module #3 ................................................................................................ 65
Test for Module #4 ................................................................................................ 67
Test for Module #5 ................................................................................................ 69
Test for Module #6 ................................................................................................ 71
Test for Module #7 ................................................................................................ 73
Test for Module #8 ................................................................................................ 75
Test for Module #9 ................................................................................................ 77
Test for Module #10 .............................................................................................. 79
Test for Module #11 .............................................................................................. 81
Test for Module #12 .............................................................................................. 83
Test for Module #13 .............................................................................................. 85
Test for Module #14 .............................................................................................. 87
Test for Module #15 .............................................................................................. 89
Test for Module #16 .............................................................................................. 91

## Answers to the Tests

Solutions to the Test for Module #1 ............ 95
Solutions to the Test for Module #2 ............ 97
Solutions to the Test for Module #3 ............ 99
Solutions to the Test for Module #4 ............ 100
Solutions to the Test for Module #5 ............ 102
Solutions to the Test for Module #6 ............ 104
Solutions to the Test for Module #7 ............ 105
Solutions to the Test for Module #8 ............ 107
Solutions to the Test for Module #9 ............ 109
Solutions to the Test for Module #10 ............ 111
Solutions to the Test for Module #11 ............ 113
Solutions to the Test for Module #12 ............ 115
Solutions to the Test for Module #13 ............ 116
Solutions to the Test for Module #14 ............ 117
Solutions to the Test for Module #15 ............ 119
Solutions to the Test for Module #16 ............ 121

# TEACHER'S NOTES
*The Human Body: Fearfully and Wonderfully Made!*
*Solutions and Tests*
Dr. Jay L. Wile

Thank you for purchasing *The Human Body: Fearfully and Wonderfully Made!* I designed this modular course specifically to meet the needs of the homeschooling parent. I am very sensitive to the fact that most homeschooling parents do not know the upper-level sciences very well, if at all. As a result, they consider it nearly impossible to teach to their children. This course has several features that make it ideal for such a parent.

1. The course is written in a conversational style. Unlike many authors, I do not get wrapped up in the desire to write formally. As a result, the text is easy to read and the student feels more like he or she is *learning*, not just reading.

2. The course is completely self-contained. Each module includes the text of the lesson, experiments to perform, questions to answer, and a test to take. The solutions to the questions are fully explained, and the test answers are provided. The experiments are written in a careful, step-by-step manner that tells the student not only what he or she should be doing, but also what he or she should be observing.

3. Most importantly, this course is Christ-centered. In every way possible, I try to make science glorify God. One of the most important things that you and your student should get out of this course is a deeper appreciation for the wonder of God's Creation!

I hope that you enjoy using this course as much as I enjoyed writing it!

## Pedagogy of the Text

(1) There are two types of exercises that the student is expected to complete: "on your own" questions, and an end-of-the module study guide.

- The "on your own" questions should be answered as the student reads the text. The act of answering these questions will cement in the student's mind the concepts he or she is trying to learn. Answers to these problems are in the student text.

- The study guide should be completed in its entirety after the student has finished each module. Answers to the study guide questions are in this book.

The student should be allowed to study the solutions to the "on your own" questions while he or she is working on them. When the student reaches the study guide, however, the solutions should be used only to check the student's completed work.

(2) In addition to the solutions to the study guides, there is a test for each module in this book, along with the answers to the test. **I strongly recommend that you administer each test once the student has completed the module and all associated exercises. The student should be allowed to have only pencil, paper, and a calculator.** I understand that many homeschoolers do not like the idea of administering tests. However, if your student is planning to attend college, it is *absolutely* necessary that he or she become comfortable with taking tests!

(3) The best way to grade the tests is to assign one point for every answer that the student must supply. Thus, if a question has three parts and an answer must be supplied for each part, the question should be worth 3 points. The student's percentage correct, then, is simply the number of answers the student got right divided by the total number of answers times 100. The student's letter grade should be based on a 90/80/70/60 scale.

(4) All definitions presented in the text are centered. The words will appear in the study guide and their definitions need to be memorized.

(5) Words that appear in bold-face type in the text are important terms that the student should know.

(6) The study guide gives your student a good feel for what I require him or her to know for the test. Any information needed to answer the study guide questions is information that the student must know for the test.

## Experiments

The experiments in this course are designed to be done as the student is reading the text. I recommend that your student keep a notebook of these experiments. This notebook serves two purposes. First, as the student writes about the experiment in the notebook, he or she will be forced to think through all of the concepts that were explored in the experiment. This will help the student cement them into his or her mind. Second, certain colleges might actually ask for some evidence that your student did, indeed, have a laboratory component to his or her science course. The notebook will not only provide such evidence but will also show the college administrator the quality of the science instruction that you provided to your student. I recommend that you perform your experiments in the following way:

- When your student gets to the experiment during the reading, have him or her read through the experiment in its entirety. This will allow the student to gain a quick understanding of what her or she is to do.

- Once the student has read the experiment, he or she should then start a new page in his or her laboratory notebook. The first page should be used to write down all of the data taken during the experiments and perform any exercises discussed in the experiment.

- When the student has finished the experiment, he or she should write a brief report in his or her notebook, right after the page where the data and exercises were written. The report should be a brief discussion of what was done and what was learned. The discussion should be written so that someone who has never read the book can read the discussion and figure out what basic procedure was followed and what was learned as a result of the experiment.

- **<u>PLEASE OBSERVE COMMON SENSE SAFETY PRECAUTIONS. The experiments are no more dangerous than most normal, household activity. Remember, however, that the vast majority of accidents do happen in the home!</u>**

# Solutions To The

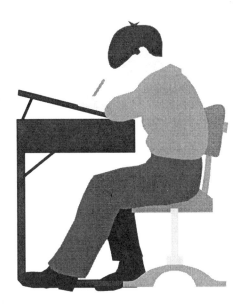

# Study Guides

# SOLUTIONS TO THE MODULE #1 STUDY GUIDE

1. a. <u>Gross anatomy</u> - The study of the macroscopic structures of an organism

   b. <u>Microscopic anatomy</u> - The study of the microscopic structures of an organism

   c. <u>Physiology</u> - The study of the functions of an organism and its parts

   d. <u>Histology</u> - The study of tissues

   e. <u>Organ</u> - A group of tissues specialized for a particular function

   f. <u>Tissues</u> - Groups of cells specialized for a particular function

   g. <u>Homeostasis</u> - A state of equilibrium in the body with respect to its functions, chemical levels, and tissues

   h. <u>Effector</u> - A structure in the body that can change the value of a variable

   i. <u>Selective permeability</u> - The ability to let certain materials in or out while restricting others

   j. <u>Endocytosis</u> - The process by which large molecules are taken into the cell

   k. <u>Exocytosis</u> - Transportation of material from inside the cell to outside the cell

2. This would be an <u>anatomy</u> course. Anatomy concentrates on the structures of an organism. Physiology studies how an organism and its parts *function*.

3. An organism is organized as follows: <u>organism, organ systems, organs, tissues, cells, organelles, and molecules</u>.

4. You would be looking at the following levels: <u>tissues, cells, and organelles</u>. These are the things that you can see with a microscope. You do not need a microscope to see a human as a whole (organism), an organ system, or an organ. When you put an organ under the microscope, however, you then see tissues, cells, and organelles. You cannot see chemicals with a 40x, 100x, 400x, 1000x microscope.

5. The four types of tissue are <u>nervous tissue, muscle tissue, connective tissue, and epithelial tissue</u>.

6. a. <u>Epithelial tissue</u> makes up the lining of many organs.
   b. Muscles are made of <u>muscle tissue</u>.
   c. Cartilage is an example of <u>connective tissue</u>, as discussed in the module.
   d. The brain, spinal cord, etc. are made of <u>nervous tissue</u>.

7. Homeostasis is threatened by <u>stress</u>.

8. A negative feedback system would cause the heart rate to <u>decrease</u>, because negative feedback systems produce effects that are the opposite of the stress. A positive feedback system would cause the heart rate to <u>increase</u>, because positive feedback systems produce effects that are the same as that of the stress.

9. <u>The nervous system</u> and <u>endocrine system</u> are the organ systems most involved with negative feedback. The nervous system sends the messages to the control center, which is also often a part of the nervous system. The endocrine system also functions as a control center.

10. a. The stress is <u>a decrease in blood glucose level</u>. Remember, stress is what takes us away from homeostasis. The drop in blood glucose level was a departure from homeostasis.

b. The control center is the <u>pancreas</u>. It is the one that monitors blood glucose level and determines whether or not something should be done.

c. The effector is the <u>liver</u>. The liver caused more glucose to be released in the blood, which was the negative response to the blood glucose level decrease.

d. <u>The endocrine system is involved</u>, because hormones were released. Those hormones caused the effector to do its job.

11. The organelles and their main function are given below:

| ORGANELLE | FUNCTION |
| --- | --- |
| Nucleus | Contains DNA |
| Plasma membrane | Holds the cell together and controls entry and exit of substances |
| Ribosomes | Synthesize proteins |
| Rough endoplasmic reticulum | Intercellular transport and synthesis of proteins |
| Smooth endoplasmic reticulum | Intercellular transport and synthesis of carbohydrates and lipids |
| Golgi apparatus | Packages chemicals for secretion |
| Secretory vesicle | Secretion |
| Lysosome | Breaks down proteins, polysaccharides, nucleic acids, and lipids |
| Mitochondria | Produce energy for the cell |
| Cilia | Movement |
| Centrioles | Spindle formation for mitosis and meiosis |

12. There are <u>three</u> nucleotides in a codon.

13. An anticodon must bind to a codon. Thus, the first nucleotide on the anticodon must be able to bind to the nucleotide on the codon. Only thymine or uracil bind to adenine. RNA does not have thymine, and an anticodon is on tRNA. Thus, it must be <u>uracil</u>. In order for a nucleotide to

have been in the codon, mRNA must have been able to bind to DNA at that point. Once again, only thymine and uracil bind to adenine. DNA does not have uracil. Thus, thymine must have been on the DNA at that point.

14. You can start with interphase if you want, although that is technically not a step in mitosis. That's the cell's "normal" state. Mitosis occurs in this order: prophase, metaphase, anaphase, telophase.

15. The "X" shape occurs only when the chromosomes have been duplicated and the duplicates have not been separated from each other. This is only during prophase and metaphase. In anaphase and telophase, the chromosomes have been separated from their duplicates.

16. The fact that phospholipids have a polar head and a nonpolar tail keeps them oriented properly. Even if disturbed, they will reorganize themselves so that the heads are pointed either into the cell or towards the outside, and the tails are pointed towards each other.

17. A glycoprotein allows for identification.

18. A receptor protein takes in messages from other cells.

19. The "fluid" is the fatty part of the membrane, which is composed of phospholipids. Mosaic refers to the fact that there are several different proteins scattered throughout.

20. A substance can dissolve through the phospholipids, it can enter through a channel protein, it can enter using a carrier protein, or it can enter through endocytosis. You could have listed charged channel proteins, but that's just a variation on channel proteins. Also, you could have listed both pinocytosis and phagocytosis, but endocytosis covers them both. That's how we can cram six methods of entering the cell down into just four.

21. a. Water is small enough to enter the cell through a channel protein.

b. A protein can only enter through pinocytosis.

c. Ions are small enough to go enter through channel proteins, but they use charged channel proteins.

d. A monosaccharide is like glucose. Thus, it will need a carrier protein because of its size.

e. An invading bacterium must be engulfed using phagocytosis.

f. A lipid is a fat, which will dissolve through the phospholipids.

22. Regardless of concentrations, pinocytosis is always an active transport process.

23. The glucose moved according to the dictates of diffusion. Thus, this is passive transport, and <u>no ATP was expended</u>.

# SOLUTIONS TO THE MODULE #2 STUDY GUIDE

1. a. <u>Exocrine glands</u> - Glands that secrete substances outward through a duct

b. <u>Endocrine glands</u> - Ductless glands that secrete hormones into the bloodstream

c. <u>Merocrine glands</u> - Exocrine glands that secrete without losing cellular material

d. <u>Apocrine glands</u> - Exocrine glands that have cytoplasm in their secretions

e. <u>Holocrine glands</u> - Exocrine glands whose secretions are made up of disintegrated cells

f. <u>Extracellular matrix</u> - The chemical substances located between connective tissue cells

g. <u>Fibroblasts</u> - Spindle-shaped cells that form connective tissue proper

h. <u>Chondrocytes</u> - Mature cartilage cells

i. <u>Stromal cells</u> - Cells that provide structure or support for parenchymal cells

j. <u>Parenchymal cells</u> - Cells that provide the actual function of the tissue

k. <u>Labile cells</u> - Cells that undergo mitosis regularly and quickly

l. <u>Stable cells</u> - Cells that do not regularly undergo mitosis but are able to if the need arises

m. <u>Permanent cells</u> - Cells that cannot undergo mitosis

2. <u>Epithelial tissue, connective tissue, nervous tissue, and muscle tissue</u>

3. <u>The height of the cells and the number of layers of cells</u> affects the distance from the free surface to the basal surface. The taller the cells and the more layers there are, the greater the distance.

4. The basement membrane is the <u>glue that the epithelial cells secrete to attach the basal surface to the tissue below.</u>

5. The basement membrane is <u>avascular</u>. Thus, the cells must get oxygen and nutrients through diffusion.

6. a. This is <u>simple squamous epithelium</u>. It can be found in <u>blood vessels and deep in the lungs</u> (you only needed to mention one of those). Its function is to <u>allow for diffusion</u>.

b. This is <u>simple cuboidal epithelium</u>. It can be found in <u>the kidneys</u>. Its function is to <u>allow for diffusion as well as absorption and secretion.</u>

c. This is <u>simple columnar epithelium</u>. It can be found in <u>the stomach and intestines</u> (you only needed to mention one of those). Its function is <u>complex absorption and secretion as well as protection</u>.

d. This is <u>stratified squamous epithelium</u>. It can be found in <u>the skin</u>. Its function is <u>to form a barrier</u>.

e. This is <u>stratified transitional epithelium</u>. It can be found in <u>the urinary bladder</u>. Its function is <u>to provide protection and the ability to stretch</u>.

f. This is <u>pseudostratified epithelium</u>. It can be found in <u>the airways of the lungs</u>. Its function is <u>to produce and move mucus</u>.

7. <u>Merocrine</u> glands work by exocytosis. Thus, no cellular material enters the secretion.

8. The four types of connective tissue are <u>connective tissue proper, bone, cartilage, and blood</u>.

9. <u>Collagen is found in all connective tissue proper</u>.

10. a. This is <u>loose connective tissue</u>. It provides for <u>light-duty binding</u>. It can be found <u>under the skin</u>.

b. This is <u>dense irregular connective tissue</u>. It provides for <u>strength in all directions</u>. It can be found in the <u>lower layer of the skin</u> (the dermis).

c. This is <u>dense regular connective tissue</u>. It provides for <u>tensile strength</u>. It can be found in the <u>ligaments and tendons</u> (you need to mention only one).

d. This is <u>adipose tissue</u>. It provides for <u>insulation and bracing organs</u>. It can be found in the <u>mammary glands, under the skin, and around the kidneys</u> (you need to mention only one).

11. <u>The ground substance in cartilage is firm. The chondrocytes need somewhere they can live. A lacuna provides that.</u>

12. a. This is <u>hyaline cartilage</u>. It can be found in <u>the bridge of the nose, the costal cartilage of the ribs, and caps on the bones of joints</u>. It functions like hard plastic, providing <u>firmness with resilience</u>.

b. This is <u>fibrocartilage</u>. It can be found in the <u>joints of the backbone</u>. It provides <u>tough binding and resilient support</u>.

c. This is <u>elastic cartilage</u>. It can be found in the <u>outer ear</u>. It provides <u>flexible support</u>.

13. <u>Serous membranes are found around organs. They lubricate organs so that they can move against one another. Mucous membranes are found in tubes that open to the outside. They provide protection. Finally, synovial membranes are found around the joints, and they provide joint lubrication.</u>

14. <u>Yes, it will be as good as new</u>. Since the parenchymal cells are the functional ones, this means that the functional cells can be replaced by mitosis.

15. Connective tissue provides support. Thus, the cells are <u>stromal</u>.

## SOLUTIONS TO THE MODULE #3 STUDY GUIDE

1. a. <u>Epidermis</u> - The outer portion of the skin, formed by epithelial tissue which rests on the dermis

b. <u>Dermis</u> - Dense irregular connective tissue that forms the deep layer of the skin

c. <u>Hypodermis</u> - Loose connective tissue underneath the dermis which connects the dermis to muscle or bone

d. <u>Hemopoiesis</u> - The process of manufacturing blood cells

e. <u>Compact bone</u> - Dense bone matrix enclosing only a few small spaces

f. <u>Cancellous bone</u> - Bone with many small spaces or cavities surrounding the bone matrix

g. <u>Ossification</u> - Bone formation

h. <u>Articular cartilage</u> - Hyaline cartilage that covers the ends of a bone in a joint

i. <u>Axial skeleton</u> – The portion of the skeleton that supports and protects the head, neck, and trunk

j. <u>Appendicular skeleton</u> – The portion of the skeleton that attaches to the axial skeleton and has the limbs attached to it

k. <u>Suture</u> - A junction between flat bones of the skull

l. <u>Process</u> - A projection on a bone

m. <u>Meatus</u> - A passageway

n. <u>Foramen</u> - A hole

o. <u>Sinus</u> - A hollowed out space in a bone

2. The two layers of skin are the <u>epidermis and the dermis. Keratinized cells are found in the epidermis, and papillae are found in the dermis.</u>

3. <u>The dermal papillae allow for maximum blood flow to the basement membrane. This helps the epidermal cells get the nutrients they need</u>. Remember, epithelial tissue is avascular. Thus, the only source of nutrients is the blood in the dermis.

4. If a cut is along the skin's lines of cleavage, there is not much tension on it. As a result, it heals quickly with little scarring. If a cut goes across the lines of cleavage, there is tension which pulls it apart, making it harder to heal and more likely to scar.

5. The hair follicle contains epithelial tissue. Thus, if the epidermis is destroyed in a second-degree burn, the epithelial tissue in the hair follicle can reproduce and make more epidermis.

6. The hypodermis always contains loose connective tissue. Most of the time, there is also adipose tissue in the hypodermis.

7. Sebaceous glands secrete oil. That oil softens the skin and protects from certain pathogens.

8. Sweat glands secrete sweat. The sweat cools the body via evaporation.

9. The epidermal layers from deep to superficial are: stratum basale, stratum spinosum, stratum granulosum, stratum lucidum, and stratum corneum.

10. Only the two deepest layers, stratum basale and stratum spinosum, contain living cells. The cells keratinize in the stratum granulosum and are, at that point, dead.

11. Melanocytes are cells that produce melanin. They give the skin its color, depending on the amount of melanin they produce. Melanin also protects the skin from ultraviolet radiation.

12. a. nerves  b. hairs  c. sweat pore  d. blood vessel loop in dermal papilla  e. sweat duct
f. sweat gland  g. vein  h. artery  i. adipose tissue  j. arrector pili  k. sebaceous gland
l. hair follicle  m. hypodermis  n. dermis  o. epidermis

13. The three sections of hair are the medulla, the cortex, and the cuticle. The medulla contains cells with soft keratin, the cortex and cuticle contain cells with hard keratin.

14. The hair matrix is a mass of undifferentiated cells. These cells reproduce, forming new hair cells.

15. Sweat glands are merocrine glands. Even apocrine sweat glands are actually merocrine glands. They were just named wrongly.

16. Sebaceous glands are holocrine glands.

17. a. Ribs are flat bones.

b. Metatarsals are longer than they are wide. Thus, they are long bones.

c. Carpals are about as long as they are wide. Thus, they are short bones.

d. The patella is an example of a sesamoid bone.

e. The coxa is an <u>irregular bone</u>.

18. In long bones, the yellow bone marrow is found in the <u>medullary cavity</u>. Red bone marrow may be found in the cancellous bone nearest the attachment to the trunk.

19. The functions of the skeletal system are: <u>support, protection, movement, storage, and hemopoiesis</u>.

20. a. <u>skull</u>  b. <u>mandible</u>  c. <u>thoracic cage</u>  d. <u>ulna</u>  e. <u>radius</u>  f. <u>carpals</u>  g. <u>metacarpals</u>  h. <u>phalanges</u>  i. <u>femur</u>  j. <u>patella</u>  k. <u>tibia</u>  l. <u>fibula</u>  m. <u>tarsals</u>  n. <u>metatarsals</u>  o. <u>phalanges</u>  p. <u>coxa</u>  q. <u>sacrum</u>  r. <u>humerus</u>  s. <u>costal cartilage</u>  t. <u>rib</u>  u. <u>sternum</u>  v. <u>cervical vertebrae</u>  w. <u>thoracic vertebrae</u>  x. <u>lumbar vertebrae</u>  y. <u>clavicle</u>  z. <u>pectoral girdle</u>  aa. <u>scapula</u>  bb. <u>pelvic girdle</u>

21. a. <u>frontal bone</u>  b. <u>ethmoid bone</u>  c. <u>lacrimal bone</u>  d. <u>zygomatic bone</u>  e. <u>vomer</u>  f. <u>maxilla</u>  g. <u>mandible</u>  h. <u>parietal bone</u>  i. <u>temporal bone</u>  j. <u>occipital bone</u>  k. <u>sphenoid bone</u>  l. <u>nasal bone</u>

# SOLUTIONS TO THE MODULE #4 STUDY GUIDE

1. a. <u>Osteoblast</u> - A bone-forming cell

b. <u>Osteocyte</u> - A mature bone cell surrounded by bone matrix

c. <u>Osteoclast</u> - A large, multinucleated cell that breaks down bone

d. <u>Hematoma</u> - A localized mass of blood that is confined to an organ or some definable space

e. <u>Callus</u> - A mass of tissue that connects the ends of a broken bone

f. <u>Anatomical position</u> - The position acquired when one stands erect with the feet facing forward, the upper limbs hanging at the sides, and the palms facing forward with the thumbs to the outside

2. <u>The two substances are collagen and a calcium/phosphorus mineral called hydroxyapatite. The collagen gives the bones some flexibility and tensile strength, and the hydroxyapatite gives the bones their hardness.</u>

3. If the cell is completely surrounded by bone matrix, it is a mature bone cell. Therefore, it is an <u>osteocyte</u>.

4. Only an <u>osteoclast</u> has more than one nucleus.

5. Compact bone tissue is made of osteons. Thus, this is <u>cancellous bone tissue</u>.

6. <u>Cancellous bone tissue</u> is made of trabeculae. You find <u>red bone marrow</u> in the spaces between trabeculae.

7. Osteons are made of layers of tissue that form cylinders. These are called <u>concentric lamellae</u>. Between the concentric lamellae, you find <u>interstitial lamellae</u>.

8. <u>Canaliculi are the extensions of osteocytes which allow the cells in bone tissue to communicate with one another.</u>

9. <u>First, all new bone tissue is cancellous bone. Some new bone tissue must be compact bone, so cancellous bone tissue often needs to be remodeled into compact bone tissue. Second, bones increase and decrease in mass based on the stress they experience.</u> In other words, bones are constantly being remodeled so that they can meet the changing demands that you place on them. <u>Third, bone is remodeled in order to re-shape the bone as needed. Fourth, bone is remodeled to repair broken bones. Fifth, bone is remodeled to replace worn collagen or hydroxyapatite. Finally, bone is remodeled to regulate the calcium levels in your body.</u>

10. The epiphyseal plate does not thicken because <u>cartilage is ossified at the same rate in which it is added</u>. As a result, the diaphysis gets bigger, but the epiphyseal plate does not.

11. The epiphyseal plate ossifies on the <u>diaphysis end</u>.

12. The bone <u>can still grow, but ONLY in width</u>. Remember, osteoblasts can still add bone tissue on top of existing bone tissue. This will make the bone THICKER. Without an epiphyseal plate, there is no way the bone can grow longer.

13. Appositional bone growth occurs when <u>osteoblasts lay new bone tissue on top of old bone tissue</u>.

14. The proper order is:

b. <u>A hematoma forms.</u>
d. <u>The callus forms.</u>
c. <u>The callus is ossified.</u>
a. <u>The external callus is removed by osteoclasts and cancellous bone is remodeled as needed.</u>

15. <u>The external callus helps hold the broken pieces of bone together. The internal callus ossifies to become the new bone tissue.</u>

16. Calcitonin is secreted by the <u>thyroid gland</u>, and PTH is secreted by the <u>parathyroid glands</u>.

17. <u>Calcitonin decreases osteoclast activity, while PTH increases osteoclast activity.</u>

18. The only way the calcitonin levels in a person's body can rise is if the person's thyroid is secreting calcitonin. That happens when <u>the blood calcium level is too high</u>, because this hormone decreases osteoclast activity. Thus, the thyroid gland senses that there is too much calcium in the blood and acts to correct that by secreting calcitonin, which decreases osteoclast activity. This lowers the amount of calcium getting into the blood from bone remodeling.

19. The <u>anterior pituitary gland</u> secretes HGH. <u>HGH stimulates bone growth by increasing osteoblast activity</u>.

20. <u>The sex hormones increase osteoblast activity, which first stimulates bone growth. However, at the same time, they stimulate ossification of the epiphyseal plates, which eventually halts bone growth, at least in terms of length.</u>

21. The three major joint types are: <u>fibrous joints, cartilaginous joints, and synovial joints. The synovial joints are responsible for most of the skeleton's movement</u>.

22. a. <u>bursa</u>   b. <u>fibrous capsule</u>   c. <u>synovial membrane</u>   d. <u>synovial fluid</u>
e. <u>articular cartilage</u>   f. <u>inner layer of the periosteum</u>   g. <u>outer layer of the periosteum</u>

23. Articular cartilage <u>cushions the ends of the bones with a "hard plastic" finish</u>. This makes the bones rub against each other more smoothly and with no bone damage.

24. Synovial fluid <u>lubricates the joint</u>. This makes motion in the joint much easier than it would otherwise be.

25. The <u>synovial membrane</u> produces synovial fluid.

26. In terms of DECREASING range of motion they are: <u>ball and socket, ellipsoid, saddle, hinge, pivot, and gliding</u>. You can also use "plane" in place of gliding.

a. This is <u>plantar flexion</u>.
b. This is <u>extension</u>, since he is straightening a joint.
c. This is <u>circumduction</u>, because the person is twirling a ball and socket joint.
d. This is <u>pronation</u>.
e. This is <u>abduction</u>, because the bones are moving away from the midline.

## SOLUTIONS TO THE MODULE #5 STUDY GUIDE

1. a. <u>Sarcomere</u> - The repeating unit of a myofibril

b. <u>Neuron</u> - The functional unit of the nervous system, a nerve cell

c. <u>Synapse</u> - The interface between a nerve cell and another cell

d. <u>Neurotransmitter</u> - A chemical released by a neuron. This chemical travels across the synaptic cleft, allowing the neuron to communicate with another cell.

e. <u>Motor unit</u> - One motor neuron and all the muscle fibers it innervates

f. <u>All-or-none law of skeletal muscle contraction</u> - An individual muscle fiber contracts with equal force in response to each action potential.

g. <u>Subthreshold stimulus</u> - A stimulus too small to create an action potential in a neuron

h. <u>Threshold stimulus</u> - A stimulus strong enough to create one action potential in a neuron

i. <u>Submaximal stimuli</u> - Stimuli of increasing strength that create more action potentials along more neurons

j. <u>Maximal stimulus</u> - A stimulus which is strong enough to create action potentials in all the motor neurons innervating a whole muscle

k. <u>Muscle tone</u> - The state of partial contraction in a muscle, even when the muscle is not being used

2. <u>Muscle tissue has contractility; it has excitability; it has extensibility; and it has elasticity.</u>

3. The three types of tissue are <u>skeletal muscle, smooth muscle, and cardiac muscle.</u>

4. a. <u>epimysium</u>   b. <u>perimysium</u>   c. <u>endomysium</u>   d. <u>muscle cell (fiber)</u>   e. <u>fascicle</u>

5. <u>Skeletal muscle cells have many nuclei rather than just one. Also, the nuclei are at the edge of the cell rather than near the center.</u>

6. a. <u>actin myofilament</u>   b. <u>myosin myofilament</u>   c. <u>Z disk</u>   d. <u>H zone</u>   e. <u>A band</u>   f. <u>I band</u>

7 a. The distance between the Z disks <u>decreases</u>, because the contraction pulls the Z disks together.

b. The length of the A band <u>remains the same</u>. Remember, the length of the myosin myofilament does not change. Thus, the A band, which is defined by the length of the myosin myofilament, stays the same.

c. The length of the I band <u>decreases</u>. The actin myofilaments are pulled towards the center of the sarcomere. This increases the amount of overlap between the actin and myosin myofilaments. The I band is defined as the region with *actin myofilament only*. Since more actin overlaps with myosin, the length of the actin myofilament only region goes down.

d. The length of the H zone <u>decreases</u>. The actin myofilaments from each side of the sarcomere are pulled closer. This reduces the length of the region in which there is only myosin, which is the definition of the H zone.

e. The length of the myosin myofilament <u>remains the same</u>. The myofilaments never change in length. The amount that they overlap changes.

f. The length of the actin myofilament <u>remains the same</u>. The myofilaments never change in length. The amount that they overlap changes.

8. The proper order is:

i. <u>An action potential travels down the axon of a motor neuron.</u>
b. <u>ACh is released from the presynaptic terminal.</u>
e. <u>ACh travels across the synaptic cleft.</u>
j. <u>ACh interacts with the muscle cell membrane to create a muscle action potential.</u>
a. <u>The action potential travels down a T-tubule.</u>
g. <u>Calcium ions are released from the sarcoplasmic reticulum.</u>
d. <u>$Ca^{2+}$ binds to troponin, exposing the active sites on the actin, allowing the myosin heads to grab onto the actin.</u>
h. <u>The power stroke.</u>
c. <u>ATP binds to the myosin heads, making them release the active sites on the actin.</u>
f. <u>The return stroke.</u>

9. If the calcium ion concentration is decreasing, that means calcium ions are heading into the sarcomere. This means the muscle is <u>starting to contract</u>, because calcium ions bind to the troponin on the actin myofilaments to begin contraction.

10. If the myosin heads have just received a boost of energy, ATP must have bound to them and then broken down into ADP and P. This means the <u>return stroke</u> is about to start. Remember, the myosin heads get "primed" with energy during the return stroke. They then use that energy on the power stroke.

11. If ADP but no P is attached, that means the myosin head has bound to the active site (that's what bumps the P off). Thus, the <u>power stroke</u> is about to happen. That will kick off the ADP.

12. a. <u>presynaptic terminal</u>  b. <u>mitochondria</u>  c. <u>synaptic vesicle</u>  d. <u>synaptic cleft</u>  e. <u>postsynaptic membrane</u>

13. <u>All of the cells in the same motor unit will contract identically at the same time.</u>  If you could see several muscle cells always contracting in unison, that is a motor unit.

14. <u>Acetylcholinesterase inactivates ACh after it has interacted with the postsynaptic membrane. If it were not for this enzyme, the muscle cell could never relax once it started contracting!</u>

15. <u>ATP provides the sarcoplasmic reticulum with energy for the active transport of calcium ions into itself.</u>  This is critical for muscle relaxation.  <u>ATP attaches to the myosin heads, making them release the active sites and giving them energy for the return stroke.</u>  Without this step, the myosin heads would never let go of the active sites, and the muscle could neither relax nor contract.

16. <u>The myosin heads must be gripping the active sites and not letting go.  This must be due to a lack of ATP in the sarcomere.</u>

17. <u>No.</u>  When a muscle cell relaxes, the myosin heads let go of the active sites.  This makes it very easy for the muscle to stretch out, but that does not automatically happen.  Something else (gravity or another muscle's contraction) must cause the relaxed muscle to extend.

18. When a motor neuron is recruited, <u>it is responding to a stimulus by sending action potentials down its axon.</u>  This stimulates contraction of the muscle fiber in its control.

19. <u>A supramaximal stimulus is being applied.</u>  This will not increase the force of the contraction, since all motor units are already "on."

20. <u>The only option left is anaerobic respiration.</u>  This is not as efficient as aerobic respiration, but it gets energy to the muscle cell quickly.

21. <u>Lactic acid will build up in this cell.</u>  It is a byproduct of anaerobic respiration.  This toxin must be eliminated from the cell.

22. The increased oxygen supply is fueling aerobic respiration.  This provides energy for the muscle fiber to <u>(1) convert lactic acid into glucose and (2) remake creatine phosphate for an energy reserve.</u>  The oxygen supply is also used by the liver cells as well, because they also convert lactic acid into glucose.

# SOLUTIONS TO THE MODULE #6 STUDY GUIDE

1. a. <u>Origin</u> - The point at which a muscle's tendon attaches to the more stationary bone

b. <u>Insertion</u> - The point at which a muscle's tendon attaches to the moveable bone

c. <u>Belly</u> - The largest part of the muscle, which actually contains the muscle cells

d. <u>Mastication</u> - The process of chewing

e. <u>Extrinsic hand muscles</u> - Muscles in the forearm which create motion in the hands

f. <u>Intrinsic hand muscles</u> - Muscles within the hand which create motion in the hands

2. Muscles that work together for the same motion are synergists. The main muscle that gets the job done is the prime mover. Thus, <u>these muscles are synergists and the gluteus maximus is the prime mover</u>.

3. In this situation, we have a <u>second-class lever</u>. Look at the flexor digitorum superficialis. The tendons connect near the ends of the fingers. When it flexes the wrist, the weight it lifts is mostly the palm of the hand. The wrist is the fulcrum. Thus, the resistance is between the effort and the fulcrum.

4. a. <u>frontalis</u>   b. <u>occipitalis</u>   c. <u>sternocleidomastoid</u>   d. <u>posterior triangle</u>   e. <u>trapezius</u>
f. <u>orbicularis oculi</u>   g. <u>zygomaticus minor</u>   h. <u>zygomaticus major</u>   i. <u>masseter</u>
j. <u>orbicularis oris</u>   k. <u>platysma</u>   l. <u>temporalis</u>   m. <u>buccinator</u>   n. <u>lateral pterygoid</u>
o. <u>medial pterygoid</u>

5. The mastication muscles are: <u>temporalis, masseter, lateral pterygoid, and medial pterygoid</u>.

6. <u>The buccinator and the orbicularis oris are called the kissing muscles</u>.

7. a. <u>pectoralis minor</u>   b. <u>external oblique</u>   c. <u>internal oblique</u>
d. <u>transversus abdominis</u>   e. <u>deltoid</u>   f. <u>pectoralis major</u>   g. <u>linea alba</u>   h. <u>rectus abdominis</u>

8. The <u>linea alba</u> is not a muscle. It is a line of connective tissue.

9. The <u>deltoid and pectoralis major</u> muscles act on the arm.

10. The <u>rectus abdominus, external oblique, and internal oblique</u> muscles act on the vertebral column.

11. a. <u>trapezius</u>   b. <u>deltoid</u>   c. <u>latissimus dorsi</u>   d. <u>rhomboideus minor</u>
e. <u>rhomboideus major</u>   f. <u>levator scapulae</u>   g. <u>supraspinatus</u>   h. <u>infraspinatus</u>   i. <u>teres minor</u>
j. <u>teres major</u>   k. <u>subscapularis</u>   l. <u>biceps brachii</u>

12. a. biceps brachii   b. supinator   c. brachioradialis   d. pronator teres
e. flexor carpi radialis   f. triceps brachii   g. extensor carpi ulnaris
h. extensor carpi radialis longus   i. extensor carpi radialis brevis   j. extensor digitorum
k. extensor retinaculum

13. The extensor retinaculum is not a muscle. Its job is to secure the extensor tendons so that they do not "bow out" when the muscles work.

14. Extrinsinc hand muscles are those muscles in the forearm that cause the motion of the hand. Intrinsic hand muscles are muscles within the hand that cause the motion of the hand.

15. a. gluteus medius   b. biceps femoris   c. vastus lateralis   d. gluteus maximus
e. adductor magnus   f. semimembranosus   g. semitendinosus   h. psoas major   i. iliacus
j. sartorius   k. rectus femoris   l. gracilis   m. adductor longus   n. vastus medialis

16. The thigh flexors are: tensor fasciae latae, iliacus, psoas major, rectus femoris, sartorius, and adductor longus. The antagonists are the extensors: gluteus maximus, semitendinosus, semimembranosus, biceps femoris, and adductor magnus.

17. The thigh abductors are: gluteus maximus, gluteus medius, gluteus minimus, tensor fasciae latae. Their antagonists are the adductors: adductor longus, adductor magnus, and gracilis.

18. The flexors are: biceps brachii, brachioradialis, and pronator teres. An antagonist works against the motion of another muscle. Thus, the antagonists will be muscles that extend the forearm. The extender is the triceps brachii.

19. The supinators are the biceps brachii and the supinator. Antagonists work towards the opposite motion, so those are the pronators. The pronators are the pronator teres and the pronator quadratus.

20. a. gastrocnemius   b. soleus   c. calcaneal tendon   d. flexor digitorum longus
e. peroneus longus   f. tibialis anterior   g. extensor digitorum longus   h. extensor retinaculum
i. patellar ligament

21. The calcaneal tendon, the patellar ligament, and the extensor retinaculum are not muscles.

# SOLUTIONS TO THE MODULE #7 STUDY GUIDE

1. a. <u>Nerves</u> - Bundles of axons and their sheaths which extend from the CNS

b. <u>Ganglia</u> - Collections of neuron cell bodies which are outside of the CNS

c. <u>Spinal nerves</u> - Nerves which originate from the spinal cord

d. <u>Cranial nerves</u> - Nerves which originate from the brain

e. <u>Afferent neurons</u> - Neurons which transmit action potentials from the sensory organs to the CNS

f. <u>Efferent neurons</u> - Neurons which transmit action potentials from the CNS to the effector organs

g. <u>Somatic motor nervous system</u> - The system that transmits action potentials from the CNS to the skeletal muscles

h. <u>Autonomic nervous system</u> - The system that transmits action potentials from the CNS to the smooth muscles, cardiac muscles, and glands

i. <u>Sympathetic division</u> - Division of the ANS that generally prepares the body for physical activity

j. <u>Parasympathetic division</u> - Division of the ANS that regulates resting and nutrition-related functions such as digestion, defecation, and urination

k. <u>Association neuron</u> - A neuron that conducts action potentials from one neuron to another neuron within the CNS

l. <u>Excitability</u> – The ability to create an action potential in response to a stimulus

m. <u>Potential difference</u> – A measure of the charge difference across the cell membrane

2. The information is traveling along <u>afferent nerves</u>, since those are the nerves which carry information from the sensory organs (your eyes) to the CNS (your brain).

3. <u>The efferent division</u> is being used, since signals are sent to effector organs which are, in this case, smooth muscle cells. The most specific you can be is that the <u>parasympathetic division of the autonomic nervous system</u> is being used. Since smooth muscles are involved, this is the autonomic nervous system, and digestion-related activities are stimulated by the parasympathetic division of the autonomic nervous system.

4. a. <u>presynaptic terminals</u>  b. <u>node of Ranvier</u>  c. <u>collateral axon</u>  d. <u>Golgi apparatuses</u>
e. <u>nucleus</u>  f. <u>nucleolus</u>  g. <u>dendrites</u>  h. <u>mitochondrion</u>  i. <u>cell body</u>  j. <u>axon hillock</u>

k. axon   l. myelin sheath

5. Oligodendrocytes: bind CNS neurons together and insulate the axons

Schwann cells: insulate PNS axons

Microglia: engage in phagocytosis to fight infections

Astrocytes: form the blood-brain barrier

Non-ciliated ependymal cells: secrete cerebrospinal fluid

Ciliated ependymal cells: move cerebrospinal fluid around so that it stays homogeneous

6. a. epineurium   b. fascicle   c. axon   d. endoneurium   e. perineurium

7. A sensory nerve carries only sensory information from a receptor to the CNS, while a motor nerve carries signals only from the CNS to effector organs such as muscles. A mixed nerve carries both. Most of the nerves in the body are mixed nerves.

8. Oligodendrocytes are neuroglia found only in the CNS. The axon will not regenerate, as it must have Schwann cells to be able to regenerate.

9. The axon will not necessarily heal. Not only must it have Schwann cells, but the axon must be pretty well aligned with its severed part in order for the Schwann cells to guide the axon to it.

10. The neuron is not at rest. If it were at rest, the sodium ions would be concentrated outside of the cell and the potassium ions would be concentrated inside the cell. That's how the resting potential is maintained. Thus, this neuron is undergoing an action potential.

11. You might be tempted to say it is at rest, but the correct answer is that you cannot tell. At that POINT on the axon, there is no action potential, because the sodium ions are on the outside and the potassium ions are on the inside. However, that is just ONE POINT on the axon. The axon could have an action potentials traveling farther along it. Thus, AT THAT POINT, the axon is at rest, but there could be activity somewhere else.

12. This is a subthreshold stimulus. It caused some sodium ions to move into the cell, but not enough to create an action potential. You could also have said that it was a local potential, because that is the kind of potential a subthreshold stimulus makes.

13. The proper order is (see Figure 7.8): (c), (a), (d), (b).

14. In (a), the sodium is rushing into the cell, making the potential difference become positive. This is depolarization. In (d), the potassium is rushing out. This brings the potential difference back to a negative value and is therefore repolarization.

15. <u>The absolute refractory period</u> will keep this from happening. The action potential can stimulate another action potential, but not in the area behind, as it is in its absolute refractory period. Thus, it can only stimulate action potentials farther down the axon.

16. <u>Myelinated axons allow saltatory transmission</u>. This is a faster way to send the action potential, as it need not travel through every part of the axon. Instead, it skips from one node of Ranvier to another.

17. <u>The difference in pain and timing is due to the signal traveling on myelinated or unmyelinated axons</u>. The sharp pain comes quickly because it travels on myelinated axons. The dull ache comes a split second later because it travels on unmyelinated axons. The size of the nerve plays a role as well, but the myelination is the big component here.

18. The difference is the <u>frequency of the action potentials</u>. The action potentials are the same, since they work on the all-or-nothing principle. It is the frequency of the action potentials which determines the strength of the signal.

19. a. <u>calcium channels</u>   b. <u>synaptic vesicle</u>   c. <u>sodium channels</u>   d. <u>presynaptic terminal</u>   e. <u>synaptic cleft</u>   f. <u>postsynaptic membrane</u>

20. If the signal needs to be exactly the same the entire way, <u>a long axon should be used</u>. Synapses regulate signals and thus change them. Axons send signals to the destination unchanged.

21. <u>An inhibitory postsynaptic potential has occurred</u>. The synapse, then, is an inhibitory synapse, and that means <u>potassium ion concentration will be higher than for the resting state outside of the membrane. The sodium ion concentration will be unchanged</u>, since inhibitory synapses control only potassium channels.

22. This is an <u>excitatory synapse</u>. An excitatory synapse regulates signals by requiring several action potentials on the presynaptic neuron to trigger one signal on the postsynaptic neuron.

23. <u>Converging circuit: many inputs are digested down to just one limited output</u>

<u>Diverging circuit:  one input creates many outputs</u>

<u>Oscillating circuit: prolongs the effect of a stimulus</u>

# SOLUTIONS TO THE MODULE #8 STUDY GUIDE

1. a. <u>Gray matter</u> – Collections of nerve cell bodies and their associated neuroglia

b. <u>White matter</u> – Bundles of parallel axons and their sheaths

c. <u>Decussation</u> – A crossing over

d. <u>Vital functions</u> – Those functions of the body necessary for life on a short-term basis

e. <u>Commissures</u> - Connections of nerve fibers which allow the two hemispheres of the brain to communicate with one another

2. <u>Hypoxia is a condition in which the brain is not getting enough oxygen due to a poor blood supply. It is dangerous because it kills neurons, which cannot replaced.</u>

3. <u>Hypoglycemia is a condition in which the glucose levels in your blood get too low. This affects the brain because the neurons need glucose. Without glucose, they cannot produce the energy they need to do their jobs.</u>

4. Inferior means "bottom" and superior means "top." From bottom to top, then, Figure 8.2 tells us that the order is:

<u>medulla, pons, midbrain, hypothalamus, thalamus</u>

5. Figure 8.2 tells us that the <u>medulla, pons, and midbrain</u> are all a part of the brainstem.

6. Figure 8.2 tells us that the <u>hypothalamus and thalamus</u> are a part of the diencephalon.

7. Most of the decussation takes place in the <u>medulla</u>.

8. The <u>medulla</u> contains nuclei which control many of the body's vital functions.

9. The <u>midbrain</u> has nuclei specifically devoted to hearing and sight.

10. The <u>pons</u> relays information from the cerebrum to the cerebellum.

11. The <u>hypothalamus</u> controls the pituitary gland.

12. The <u>thalamus</u> performs crude interpretation of sensory information and then forwards it to the sensory cortex.

13. <u>The gyri are the hills and the sulci are the valleys.</u> Remember, a sulcus is a groove that separates gyri.

14. The cerebellum deals with the subconscious motor functions.

15. The corpus callosum allows the two hemispheres of the brain to communicate with one another. The general term for such structures is commissures.

16. a. brainstem   b. temporal lobe   c. lateral fissure   d. frontal lobe   e. central sulcus
f. parietal lobe   g. occipital lobe   h. cerebellum

17. a. primary somatic sensory area: receives and localizes general sensations from the entire body

b. somatic sensory association area: interprets the sensory information and puts it into context with your past experiences

c. visual association area: recognizes the meaning of visual information by putting it into context with your past experiences

d. visual cortex: interprets the basic visual information such as shape and color

e. Wernicke's area: deals with the comprehension of speech

f. auditory association area: interprets the meaning of sound by placing it into context with your past experiences

g. primary auditory area: interprets the basics of sound such as pitch and volume

h. Broca's area: initiates the muscle movements for speech

i. taste area: interprets taste

j. prefrontal area: site of motivation and foresight; regulates mood and emotion

k. premotor area: works out the sequence of signals needed for complex motion

l. primary motor cortex: controls skeletal muscle movements

18. The majority of CSF is made in the lateral ventricles.

19. The rest of the CSF is made in the third and fourth ventricles.

20. CSF cushions and protects the brain, and it also provides some nutrition.

21. The three layers which protect the brain are the dura mater, the arachnoid mater, and the pia mater. Collectively, they are called the meninges.

22. Arachnoid granulates are extensions of arachnoid mater which return CSF into the dural sinus so that it returns to the blood within the dural sinus.

23. a. dorsal root ganglion   b. afferent neuron   c. efferent neuron   d. association neuron
e. ventral root   f. gray matter (ventral horn)   g. white matter (dorsal column)

24. The reflex arc has an afferent neuron, an association neuron, and an efferent neuron. They are activated in that order. First, the afferent neuron sends the sensory information to the spinal cord, the association neuron then routes the signal to the appropriate efferent neuron, and then the efferent neuron stimulates the appropriate muscle to contract.

25. The association neuron is in the spinal cord.

26. The afferent neuron forms a diverging circuit to send information along the reflex arc and also to the brain.

27. There is a converging circuit on the efferent neuron, since both the reflex arc and the brain must control the muscle.

# SOLUTIONS TO THE STUDY GUIDE FOR MODULE #9

1. a. <u>Sensory Receptor</u> - An organ which responds to a specific type of stimulus by ultimately triggering an action potential on a sensory neuron

b. <u>Somatic receptors</u> - Sensory receptors in the skin, muscle, and tendons

c. <u>Visceral receptors</u> - Sensory receptors in the internal organs

d. <u>Special Receptors</u> - Sensory receptors in specific locations

e. <u>Mechanoreceptors</u> - Sensory receptors which respond to movement

f. <u>Thermoreceptors</u> - Sensory receptors which respond to heat or cold

g. <u>Photoreceptors</u> - Sensory receptors which respond to light

h. <u>Chemoreceptors</u> - Sensory receptors which respond to chemicals

i. <u>Nociceptors</u> - Sensory receptors which respond to pain or excess stimulation

j. <u>Cutaneous receptors</u> - Receptors in the skin

k. <u>Proprioceptors</u> - Receptors in the muscles and tendons

2. This is a <u>somatic motor neuron</u>, as autonomic neurons all have a synapse at an autonomic ganglion in between the spinal cord and the effector.

3. a. <u>In the sympathetic division, the preganglionic neuron is short and the postganglionic neuron is long. In the parasympathetic division, on the other hand, the preganglionic neuron is long and the postganglionic neuron is short</u>.

b. <u>The sympathetic division neurons can be found all over the body, but the parasympathetic nerves can only be found in the head and trunk.</u>

c. <u>The autonomic ganglia are close to the spinal cord in the sympathetic division and close to the effector in the parasympathetic division</u>.

4. a. Since hair follicle receptors are in the skin, they are <u>somatic receptors</u>. Since they detect movement of the hair, they are <u>mechanoreceptors</u>.

b. Since olfactory neurons are associated with the special sense of smell, they are <u>special receptors</u>. They respond to chemicals, so they are <u>chemoreceptors</u>.

c. Since taste buds are associated with the special sense of taste, they are <u>special receptors</u>. They respond to chemicals, so they are <u>chemoreceptors</u>.

d. Since pain receptors in the kidney are on an organ, they are <u>visceral receptors</u>. Since they are pain receptors, they are <u>nociceptors</u>.

e. Since the free nerve endings that detect cold temperatures are in the skin, they are <u>somatic receptors</u>. Since they detect cold, they are <u>thermoreceptors</u>.

f. Since rods and cones are associated with the special sense of sight, they are <u>special receptors</u>. Since they detect light, they are <u>photoreceptors</u>.

g. Since hair cells in the macula of the vestibule are associated with the special sense of balance, they are <u>special receptors</u>. Since they detect movement, they are <u>mechanoreceptors</u>.

h. Since the Golgi tendon organ is in the tendon, it is a <u>somatic receptor</u>. It detects movement, so it is a <u>mechanoreceptor</u>.

5. Projection tells you the location. Thus, <u>projection tells you it is in the back of the head</u>. Modality tells you whether the touch is soft, pleasant, sharp, painful, etc. Thus, <u>modality tells you that it is painful</u>.

6 a. <u>Free nerve endings: receptors for heat, cold, movement, itch, and pain</u>

b. <u>Merkel's disks: receptors for light touch</u>

c. <u>Hair follicle receptors: receptors that detect the movement of hair</u>

d. <u>Pacinian corpuscle: pressure receptors</u>

e. <u>Meissner's corpuscles: two-point discrimination</u>

f. <u>Ruffini's organ: pressure and stretch receptors</u>

7. <u>Muscle spindle detects the extent to which the muscle is relaxed, while Golgi tendon organs detect the extent to which the muscles are contracted</u>.

8. The substance must reach the olfactory hairs and be recognized by them. Thus, these conditions must be met:

First, the substance must get airborne. Thus, <u>the substance must be volatile</u>.

Second, the substance must get to the olfactory epithelium. Thus, <u>the airborne chemical must rise into the olfactory recess</u>.

Third, the substance must get through the mucous layer and into the cell. Thus, the substance must be at least somewhat water soluble and somewhat lipid soluble.

Finally, the substance must be recognized by the receptors. Thus, the substance must be able to bind to a receptor on the olfactory neuron.

9. The residents don't recognize the smell because olfactory receptors are quick to adapt. Thus, after a while, they do not send the signals to the brain. You will stop smelling it eventually, since your olfactory receptors will adapt, too.

10. The tongue contains circumvallate papillae, filiform papillae, foliate papillae, and fungiform papillae. All but filiform papillae have taste buds.

11. To fully taste a substance, you need to excite all of the taste buds for the different kinds of tastes (salty, sour, bitter, and sweet). Since the taste buds which respond to the different tastes are located on different parts of the tongue, the substance must be spread all over the tongue to excite all of them.

12. a. auricle  b. external auditory meatus  c. tympanic membrane  d. auditory ossicles
e. semicircular canals  f. cochlea

13. a. ultricular macula  b. ampulla  c. vestibule (made of utricle and saccule)
d. saccular macula  e. cochlea  f. otoliths  g. gelatinous matrix  h. kinocilium
i. stereocilia  j. hair cell  k. support cell  l. cupula  m. hair cell
n. crista ampullaris

14. Structures a, c-d, and f-k are involved in static equilibrium, while b, and l-n are involved in dynamic equilibrium. The cochlea is used in hearing, not balance.

15. The sound starts at the auricle and heads to the ear drum. That vibrates the ossicles, and they vibrate the oval window, which vibrates the perilymph. The perilymph vibrates the basilar membrane, which vibrates the endolymph, causing the tectorial membrane to vibrate. The order, then, is:

tympanic membrane, malleus, incus, stapes, perilymph, basilar membrane, endolymph, tectorial membrane

16. a. lens: bends light to focus it on the retina

b. sclera: maintains the shape of the eye, protects the inner components of the eye, and provides a point of attachment for the muscles that move the eye

c. optic nerve: carries action potentials to the brain

d. vitreous humor: gives the general shape to the eyeball by inflating it

e. retina: contains the light receptors that detect light

f. choroid: supplies the eye's tissues with oxygen and nutrients

g. conjunctiva: protects and lubricates the eye

h. posterior chamber: holds the aqueous humor

i. anterior chamber: holds the aqueous humor

j. cornea: covers the eye and bends light for focusing

k. pupil: allows light to enter the eye

l. iris: controls the size of the pupil

m. suspensory ligaments: connects the ciliary body to the lens

n. ciliary body: contains the ciliary muscle which changes the shape of the lens

17. Cones detect color, while rods detect low levels of light.

18. Cones are concentrated in the fovea centralis

19. Accommodation is the process of the lens changing shape to adjust the eye's focus at different distances.

# SOLUTIONS TO THE STUDY GUIDE FOR MODULE #10

1. a. <u>Neurosecretory cells</u> - Neurons of the hypothalamus that secrete neurohormone rather than neurotransmitter

b. <u>Prostaglandins</u> - Biologically active lipids which produce many effects in the body, including smooth muscle contractions, inflammation, and pain

2. a. <u>The nervous system responds much more quickly than the endocrine system.</u>

b. <u>The influence of the endocrine system lasts longer than that of the nervous system.</u>

c. <u>The nervous system controls muscles and glands, while the endocrine system controls virtually every cell in the body.</u>

d. <u>In the nervous system, the signal strength is determined by the frequency of the signal. In the endocrine system, it is determined by the amount of hormone released.</u>

e. <u>The nervous system is more difficult to repair than the endocrine system.</u>

3. Hormones must be carried in the blood. Thus, <u>if the hormone cannot dissolve in water, it will need a carrier protein</u> in order to be transported by the blood, which is mostly water.

4. <u>Hormones are eliminated by the kidneys in the urine and by the liver in the feces.</u>

5. a. <u>hypothalamus</u>  b. <u>pituitary</u>  c. <u>thyroid</u>  d. <u>adrenals</u>  e. <u>ovaries in females</u>
f. <u>pineal body</u>  g. <u>parathyroids</u>  h. <u>thymus</u>  i. <u>pancreas</u>  j. <u>testes in males</u>

6.

| Endocrine Gland | Hormone Produced | Hormone Function |
|---|---|---|
| Hypothalamus | Growth hormone releasing hormone (GH-RH) | Increases the release of GH from the anterior pituitary |
|  | Corticotropin releasing hormone (CRH) | Increases the release of ACTH from the anterior pituitary. |
|  | Gonadotropin releasing hormone (GnRH) | Increases the release of FSH and LH from the anterior pituitary |
|  | Thyroid stimulating hormone releasing hormone (TSH-RH) | Increases the release of TSH from the anterior pituitary |
|  | Prolactin inhibiting hormone (PIH) | Decreases the release of PRL from the anterior pituitary |
| Anterior Pituitary | Growth hormone (GH) | Increases growth in most tissues |
|  | Thyroid stimulating hormone (TSH) | Increases the release of thyroxin from the thyroid gland |
|  | Adrenocorticotropic hormone (ACTH) | Increases the release of cortisol from the adrenal cortex |

| | | |
|---|---|---|
| Anterior Pituitary | Luteinizing hormone (LH) | Stimulates ovaries or testes |
| | Follicle stimulating hormone (FSH) | Stimulates ovaries or testes |
| | Prolactin (PRL) | Stimulates milk production in the breasts |
| | Melanocyte stimulating hormone (MSH) | Increases the synthesis of melanin in melanocytes |
| Posterior Pituitary | Antidiuretic hormone (ADH) | Increases the retention of water by the kidneys |
| | Oxytocin (OT) | Increases the contractions of the uterus during birth and promotes the release of breast milk. |
| Thyroid | Thyroxin (TH) | Increases the metabolic rate of most cells |
| | Calcitonin | Inhibits osteoclast activity, lowering blood calcium levels |
| Parathyroids | Parathyroid hormone (PTH) | Increases blood calcium levels by increasing osteoclast activity |
| Adrenal medulla | Epinephrine (E) | Enhances sympathetic response |
| | Norepinephrine (NE) | Enhances sympathetic response |
| Adrenal cortex | Cortisol | Increases protein and fat breakdown in most tissues |
| | Aldosterone | Increases the retention of sodium by the kidneys |
| Pancreas | Insulin | Lowers blood glucose by stimulating cells to absorb glucose |
| | Glucagon | Raises blood glucose by causing liver to release glucose |
| Ovaries | Estrogen | Sex hormone in females |
| | Progesterone | Sex hormone in females |
| Testes | Testosterone | Sex hormone in males |
| Pineal body | Melatonin | Affects release of GnRH by hypothalamus |
| Thymus | Thymosin | Develops immune functions |

7. The three types of hormones are <u>amines, steroids, and peptide/proteins</u>. <u>The peptide/proteins stimulate membrane-bound receptors only</u>, because they are too big to get into the cell. <u>Steroids stimulate intracellular receptors only</u>, because they are fat-soluble and can diffuse right through the cell membrane. <u>Amines stimulate both</u>, depending on the particular hormone.

8. **<u>Nonhormonal control</u>**: <u>The level of a chemical other than a hormone affects an endocrine gland. Variations of that chemical level will stimulate or inhibit hormone secretion.</u>

**Direct neural control**: The nervous system innervates the gland with neurons. Those neurons secrete neurotransmitter to stimulate the gland to secrete the hormone, or the neurons secrete the hormone directly as a neurohormone.

**Hormone control**: One gland releases a hormone which will stimulate another gland to release a different hormone.

9. Since the hormone appears in a repeating pattern that can be predicted, it is not acute response. Since the variation is strong, it is not constant secretion. Thus, this must be cyclic secretion.

10. This is an intracellular receptor. If the protein has to be made, the receptor must have activated a gene in the nucleus.

## SOLUTIONS TO THE STUDY GUIDE FOR MODULE #11

1 a. <u>Viscosity</u> - The resistance to flow and alteration of shape due to cohesion

b. <u>Plasma</u> - The fluid portion of the blood, which is mostly water

c. <u>Formed elements of blood</u> - The cells and cell parts of blood produced by the bone marrow

d. <u>Erythrocytes</u> - Red blood cells which carry the oxygen in blood

e. <u>Leukocytes</u> - White blood cells which perform various defensive functions in blood

f. <u>Platelets</u> - Cell fragments in blood which help prevent blood loss

g. <u>Diapedesis</u> - Passage of any formed element of blood through the blood vessel and into the tissue spaces

h. <u>Chemotaxis</u> - Attraction of cells to chemical stimuli

i. <u>Hemopoiesis</u> - The process by which the formed elements of blood are made in the body

j. <u>Hemostasis</u> - The process by which the body stops blood loss

k. <u>Coagulation factors</u> - Proteins in blood plasma which help initiate the blood clotting process

l. <u>Antigen</u> - A protein or carbohydrate that, when introduced in the blood, triggers the production of an antibody

m. <u>Arteries</u> - Blood vessels that carry blood away from the heart

n. <u>Capillaries</u> - Tiny, thin-walled blood vessels that allow the exchange of gases and nutrients between the blood and cells

o. <u>Veins</u> - Blood vessels that carry blood back to the heart

p. <u>Pulmonary circulation</u> - Circulation of the blood over the air sacs of the lungs

q. <u>Systemic circulation</u> - Circulation of the blood through the other tissues of the body

r. <u>Systolic phase</u> - The phase of the cardiac cycle in which the ventricles contract

s. <u>Diastolic phase</u> - The phase of the cardiac cycle in which the ventricles relax

t. <u>Cardiac cycle</u> - One complete round of systole and diastole

u. <u>Arterioles - The smallest arteries that still have three tunics</u>

v. <u>Venules - Small veins that do not have three tunics but instead have only an endothelium, a basement membrane, and a few smooth muscle cells</u>

2. Blood is more dense than water. Thus, <u>the balloon would sink</u>.

3. <u>The pH of blood ranges from 7.35 to 7.45. It has to be tightly controlled, because many of the chemical reactions which control the body work properly only in a narrow range of pH.</u>

4. <u>Blood is made of plasma (55%) and formed elements (45%).</u>

5. <u>50% of blood is water.</u>

6. <u>Plasma is mostly water. It also contains proteins, ions, nutrients, gases, regulatory chemicals, and waste.</u>

7. <u>Erythrocytes make up most of the blood's formed elements. There are also leukocytes and blood platelets.</u>

8. <u>Hemoglobin carries oxygen in the blood.</u> It also carries some carbon dioxide, but not much.

9. <u>Iron must be present</u>, as that is the site at which oxygen molecules bind to the hemoglobin.

10. <u>Red blood cells have no nucleus</u>, once they are mature. They therefore cannot make the proteins that they need. Thus, they cannot repair damage or replace degenerated proteins.

11. <u>The granulocytes are neutrophils, basophils, eosinophils. The agranulocytes are lymphocytes and monocytes.</u>

12. <u>Neutrophils fight infections by phagocytosis.</u>
    <u>Basophils release histamine and heparin.</u>
    <u>Eosinophils are anti-inflammatory.</u>
    <u>Lymphocytes produce antibodies.</u>
    <u>Monocytes fight infections by phagocytosis.</u>

13. <u>Blood cells are formed from stem cells found in bone marrow.</u>

14. <u>Hemostasis involves vasoconstriction, platelet plug formation, and blood coagulation.</u>

15. <u>A thrombus is the plug formed in platelet plug formation.</u>

16. <u>In stage 1, prothrombinase is made. In stage two, thrombin is made. In stage 3, fibrin is made.</u>

17. <u>Coagulation factors play a critical role in stage 1.</u>

18. <u>If a blood coagulation factor is present, then coagulation may or may not be occurring. If an activated factor is present, coagulation is occurring, because factors are only activated during the coagulation process.</u>

19. <u>Type O- is the universal donor.</u> It has no antigens at all. Thus, regardless of the antibodies in the recipient's blood, there will be no substantial reaction, except for the possibility of a reaction between the *donor's* antibodies and the recipient's blood.

20. <u>Type AB+ is the universal recipient.</u> It has no antibodies. Thus, regardless of the antigens in the donor's blood, the recipient has no antibodies to attack the cells. Of course, the *donor's* antibodies can attack the recipient's cells, but the donor's antibodies are diluted. They don't last long, either.

21. A and B are dominant over O and co-dominant with each other. Since A and B are dominant over O, the parents could each have an O allele that is just not expressed. <u>Thus, the child could be type A, type B, type AB, or type O.</u> If the mother gives an O and the father gives an A, for example, the child is A. If the father gives an O and the mother gives a B, the type is B. If the mother gives an O and the father gives an O, the type is O. If the father gives an A and the mother gives a B, the type is AB.

22. Rh-positive is dominant. <u>Thus, they each must have a Rh-negative allele.</u> That way, they each donated the recessive allele to the child, and the child was Rh-negative as a result.

23. <u>Blood travels into the right atrium from the vena cava. It then goes to the right ventricle and leaves the heart through the pulmonary trunk to the lungs. It travels through the pulmonary arteries, into arterioles, and then into the capillaries of the lungs. There, it picks up oxygen. The blood then flows into venules, veins, and back to the heart through the pulmonary veins. It enters the heart in the left atrium, travels to the left ventricle, and then leaves the heart through the aorta. It then travels through arteries, arterioles, and finally capillaries. At that point, it gives oxygen to the tissues and travels back to the heart in venules, veins, and finally the vena cava.</u>

24. Systolic refers to ventricular contraction. That's when the blood pressure in the aorta is highest. Diastolic refers to ventricular relaxation, when blood pressure in the aorta is the lowest. Thus, <u>120 is the systolic and 80 is the diastolic.</u>

25. <u>The sinoatrial node is a clump of cardiac tissue that generates action potentials which cause atrial contraction. The atrioventricular node is a clump of cardiac tissue that generates action potentials which cause ventricular contraction. The sinoatrial node is the pacemaker</u>, since it starts the process and drives the AV node at its "pace."

26. The blood pressure is lowest in the <u>veins</u>.

27. This is especially true in the veins, because skeletal muscle contraction helps pump blood back to the heart in the veins.

28. a. superior vena cava   b. right pulmonary arteries   c. right pulmonary veins
d. pulmonary trunk   e. right auricle   f. right atrium   g. right ventricle
h. inferior vena cava   i. aorta   j. left pulmonary arteries   k. left pulmonary veins
l. left ventricle   m. right atrium   n. left atrium

29. a. superior vena cava   b. pulmonary semilunar valve   c. right atrium
d. right atrioventricular canal   e. right atrioventricular valve or tricuspid valve
f. chordae tendineae   g. right ventricle   h. papillary muscles   i. aorta
j. pulmonary trunk   k. left atrium   l. left atrioventricular canal
m. left atrioventricular valve or bicuspid valve   n. aortic semilunar valve   o. left ventricle
p. interventricular septum

30. Once it leaves the lungs, it travels back to the heart via the pulmonary veins and enters the left atrium. It then passes through the left atrioventricular canal and the left atrioventricular valve (the order between these two is not important). It then enters the left ventricle. It passes through the aortic semilunar valve and out the aorta. It then goes to the body's tissues and comes back, perhaps through the superior vena cava. If you didn't list that, don't worry. It could have come back through the inferior vena cava or the coronary sinus. It then enters the right atrium, passes through the right atrioventricular canal and right atrioventricular valve (the order between these two is not important), and into the right ventricle. It then passes through the pulmonary semilunar valve and into the pulmonary trunk, where it travels to the lungs again.

# SOLUTIONS TO THE STUDY GUIDE FOR MODULE #12

1. a. <u>Lymph tissue</u> – Groups of lymphocytes and other cells which support the lymphocytes

b. <u>Lymph nodes</u> – Encapsulated masses of lymph tissue found along lymph vessels

c. <u>Lymph</u> – Watery liquid formed from interstitial fluid and found in lymph vessels

d. <u>Edema</u> – A buildup of excess fluid in the tissues, which can lead to swelling

e. <u>Immunological defense</u> – The process by which the body protects itself from pathogenic invaders such as bacteria, fungi, parasites, and foreign substances

f. <u>Diffuse lymphatic tissue</u> – Concentrations of lymphatic tissue with no clear boundaries

g. <u>Lymph nodules</u> – Lymphatic tissue arranged into compact, somewhat spherical structures

h. <u>Innate immunity</u> - An immune response that is the same regardless of the pathogen or toxin encountered

i. <u>Acquired immunity</u> - An immune response targeted at a specific pathogen or toxin

j. <u>Complement</u> - A series of 20 plasma proteins activated by foreign cells or antibodies to those cells. They (1) lyse bacteria, (2) promote phagocytosis, and (3) promote inflammation.

k. <u>Interferon</u> - Proteins secreted by cells infected with a virus. These proteins stimulate nearby cells to produce virus-fighting substances.

l. <u>Pyrogens</u> - Chemicals which promote fever by acting on the hypothalamus

m. <u>Humoral immunity</u> - Immunity which comes from antibodies in blood plasma

n. <u>Cell-mediated immunity</u> - Immunity which comes from the actions of T-lymphocytes

2. Interstitial fluid becomes lymph as soon as it enters the lymph vessels. Nothing else changes. Thus, <u>the fluid entered a lymph vessel</u>.

3. <u>Interstitial fluid enters a lymph capillary through pores in between the overlapping cells in the lymph capillary. It is pushed through the lymph vessels by contractions of the skeletal muscles, contractions of smooth muscles in the larger lymph vessels, and by slightly lower pressure in the thoracic cavity.</u>

4. The three basic functions are: <u>fluid balance, fat absorption, and immunological defense</u>.

5. Lymph is not the same throughout the lymphatic system. Lymph vessels near the intestine absorb fats, giving the lymph a milky appearance there.

6. Tonsils are groups of lymph nodules found in the throat and on the back of the tongue.

7. Peyer's patches are groups of lymph nodules found in the wall of the small intestine.

8. Afferent lymph vessels carry lymph *into* the lymph node. Efferent lymph vessels carry lymph *away from* the lymph node. There are usually more afferent vessels than efferent vessels.

9. The lymph node functions are: (1) testing the lymph for foreign invaders, (2) adding lymphocytes to the blood, and (3) filtering the lymph with macrophages.

10. The three functions of the spleen are: (1) to cleanse the blood of foreign invaders, (2) to dispose of worn-out erythrocytes, and (3) to be a reservoir of oxygen-rich blood.

11. The thymus gland (1) is the place where T-lymphocytes mature and (2) secretes thymosin.

12. Skin, mucus, vasodilation, interferon, natural killer cells, and urine give us our innate immunity. T-cells, B-cells, and antibodies provide us with acquired immunity.

13. a. one-way valve   b. afferent lymph vessels   c. germinal center   d. lymph nodule
    e. capsule   f. trabeculae   g. efferent lymph vessel   h. reticular fibers

14. a. provides a barrier against infection
    b. wash the skin and lower the pH
    c. contain antibacterial agents
    d. traps foreign materials
    e. kills pathogens in stomach
    f. contain lysozyme, which kills foreign cells
    g. washes out the urinary tract
    h. "squeeze out" populations of pathogens and secrete lactic acid
    i. kills bacteria
    j. signals other cells to defend against viral infections
    k. engulf foreign substances
    l. promote inflammation
    m. reduce inflammation to keep it in check
    n. increases blood flow and allows monocytes easier access to the tissues
    o. increase body temperature by affecting the hypothalamus

15. Antibodies are shaped like a "Y."

16. The constant region determines what group an antibody belongs in.

17. Antibodies can inactivate antigens by binding to them.

Antibodies can inactivate antigens by binding them together in groups.
Antibodies can inactivate antigens by activating complement.
Antibodies can inactivate antigens by stimulating phagocytosis.
Antibodies can inactivate antigens by stimulating inflammation.

18. Plasma B-cells release antibodies, while memory B-cells remember the infection so that they can respond to the next infection. Memory B-cells live longest.

19. There are 20 proteins on the cell membrane of every cell. These proteins make up the MHC, which is essentially unique for each individual.

20. Cytotoxic T-cells attack and lyse foreign cells, memory T-cells remember the infection for the next time, and helper T-cells promote the proliferation of cytotoxic T-cells and B-cells.

# SOLUTIONS TO THE STUDY GUIDE FOR MODULE #13

1. a. <u>Mastication</u> - The process of chewing

b. <u>Digestion</u> - The breakdown of food molecules into their individual components

c. <u>Deglutition</u> - The act of swallowing

d. <u>Peristalsis</u> - The process of contraction and relaxation of circular smooth muscles which pushes food through the alimentary canal

e. <u>Gastric juice</u> - The acidic secretion of the stomach

f. <u>Adventitia</u> - A thin layer of loose connective tissue that binds an organ to surrounding tissues or organs

g. <u>Lumen</u> - The hole in the center of a tube

h. <u>Macronutrients</u> – The nutrients the body needs in large amounts: carbohydrates, fats, and proteins

i. <u>Micronutrients</u> – The nutrients the body needs in small amounts, such as vitamins and minerals

2. a. <u>tongue</u>  b. <u>larynx</u>  c. <u>trachea</u>  d. <u>pancreas</u>  e. <u>stomach</u>  f. <u>small intestine</u>
g. <u>anus</u>  h. <u>parotid salivary gland</u>  i. <u>sublingual salivary gland</u>
j. <u>submandibular salivary gland</u>  k. <u>pharynx</u>  l. <u>esophagus</u>  m. <u>lungs</u>  n. <u>liver</u>
o. <u>gall bladder</u>  p. <u>transverse colon</u>  q. <u>ascending colon</u>  r. <u>descending colon</u>  s. <u>cecum</u>
t. <u>vermiform appendix</u>  u. <u>rectum</u>  v. <u>large intestine</u>

3. a. <u>The tongue is a part of the digestive system and is a part of the alimentary canal. It moves the food around in the mouth to form the bolus. It also provides a sense of taste.</u> If you said it was not a part of the alimentary canal, that's fine. It really depends on the way you look at things.

b. <u>The larynx is not a part of the digestive system.</u>

c. <u>The trachea is not a part of the digestive system.</u>

d. <u>The pancreas is a part of the digestive system but not a part of the alimentary canal. It makes several digestive enzymes as well as a base to neutralize the chyme.</u>

e. <u>The stomach is a part of the digestive system and the alimentary canal. It churns and mixes the food with gastric juices. The gastric juices contain stomach acid which destroys bacteria that might have been eaten with the food, and it helps dissolve the food. The gastric juices also</u>

contain some digestive enzymes that start the chemical digestion of proteins. This turns the ingested food into chyme.

f. The small intestine is a part of the digestive system and the alimentary canal. It chemically digests the food and allows the nutrients to be absorbed through its lining.

g. The rectum is a part of the digestive system and the alimentary canal. It stores feces and then forces them out of the body through the anus.

h. The salivary glands are a part of the digestive system. The are not a part of the alimentary canal. They put saliva in the mouth. The saliva partially digests starch, but it also lubricates the food and protects the mouth through its lysozyme and antibodies.

i. The pharynx is a part of the digestive system and the alimentary canal. It pushes food into the esophagus.

j. The esophagus is a part of the digestive system and the alimentary canal. It pushes food down into the stomach.

k. The lungs are not a part of the digestive system.

l. The liver is a part of the digestive system but not a part of the alimentary canal. It has many functions. The digestion-related function is to make bile. The other functions are nutrient interconversion, nutrient storage, synthesis, phagocytosis, and detoxification.

m. The gall bladder is a part of the digestive system but not a part of the alimentary canal. It concentrates bile and squirts the bile into the chyme as the chyme enters the small intestine. Bile is a chemical that aids in the digestion of fats.

n. The large intestine is a part of the digestive system and the alimentary canal. It consolidates undigested food, absorbs water from it, and turns the resulting waste into feces. Bacteria in the large intestine also produce beneficial vitamins.

o. The appendix is a part of the digestive system. It is not a part of the alimentary canal. You might have thought that it was since it is attached to the large intestine, but it is not. It sort of "hangs" off the large intestine and food never passes through it. Its function is not completely understood. Many think that it is a disease-fighting organ.

p. The anus is a part of the digestive system and the alimentary canal. It is the opening through which feces exit.

4. The epiglottis covers the larynx when you swallow. It stays open when you breathe.

5. Amylase can be found in saliva. It helps break polysaccharides down into smaller carbohydrates.

6. The soft palate rises and seals off the nasal cavity during deglutition.

7. From the outside to the inside you find the serosa or adventitia, the longitudinal muscularis, the circular muscularis, the submucosa, and the mucosa. The stomach has an extra layer: the oblique muscularis, which is deep to the circular muscularis.

8. The enzyme pepsin is secreted by the stomach. It breaks down proteins into peptides. Actually, pepsinogen is secreted by the stomach. It does not become pepsin until it hits the acid of the gastric juice.

9. The gastric glands produce mucus, which coats the stomach lining and protects it from its own gastric juice.

10. Intrinsic factor allows for the absorption of vitamin $B_{12}$ by the small intestine.

11. The stomach can absorb water, alcohol, and aspirin.

12. Gastrin increases lower esophageal sphincter tone, decreases pyloric sphincter tone, increases the rate of secretion from the gastric pits, and increases the rate of mixing waves of the stomach.

13. In order from the stomach to the small intestine: duodenum, jejunum, and ileum. The duodenum is the smallest portion.

14. Circular folds and intestinal villi increase the surface area of the small intestine. Also, the cells which absorb the nutrients are covered with microvilli, which increase their surface area.

15. Secretin: Reduces gastric juice production and increases secretion of the acid-neutralizing juice from the pancreas, which increases the pH of the chyme.

Cholecystokinin (CCK): Causes the gall bladder to contract. It also inhibits gastric secretion a bit, but not nearly as much as secretin does. It also increases the secretion of digestive enzymes from the pancreas

Gastric inhibitory peptide (GIP): Decreases the rate at which the stomach empties.

16. Maltase: Breaks down maltose.
Sucrase: Breaks down sucrose.
Lactase: Breaks down lactose.
Peptidase: Breaks down peptides.
Enterokinase: Activates trypsinogen into trypsin.

17. A cell contains lots of proteins which are vital to its functions. Producing an active protein-digesting enzyme would kill the cell.

18. The bacteria produce vitamin K, biotin, and folic acid, which are absorbed in the large intestine. They also break down other chemicals that end up in the feces.

19. One way that chyme moves through the large intestine to the rectum is via mass movements. Mass movements are triggered by food entering the stomach. Thus, the act of filling the stomach triggers mass movements, which rapidly move chyme towards the rectum.

20. Blood travels from the portal triads to the central vein. Bile flows the opposite way.

21. Bile is *not* a digestive enzyme. It emulsifies fats, making them easier to digest.

22. The salivary glands and the pancreas produce amylase.

23. The four secretions which reduce the acidity of chyme are: alkaline mucus from the duodenal glands, intestinal juice from the intestinal glands, bile from the gall bladder, and bicarbonate in pancreatic juice.

24. Vitamins A, D, E, and K are fat-soluble.

25. Vitamins typically regulate the chemical processes in the body.

26. Vitamins D and K. Vitamin D is made from sunlight hitting the skin, and vitamin K is made by bacteria in the large intestine. It is still beneficial, however, to get these vitamins in your diet as well.

# SOLUTIONS TO THE STUDY GUIDE FOR MODULE #14

1. a. <u>Upper respiratory tract</u> – The part of the respiratory system containing the nasal cavity, paranasal sinuses, and pharynx

b. <u>Lower respiratory tract</u> – The part of the respiratory system containing the larynx, trachea, bronchi, and lungs

c. <u>Ventilation</u> – The process of getting air into the lungs and getting it back out

d. <u>External respiration</u> – The process of $O_2$ and $CO_2$ exchange between the alveoli and the blood

e. <u>Internal respiration</u> – The process of $O_2$ and $CO_2$ exchange between the cells and the blood

f. <u>Pneumothorax</u> - Air in the pleural cavity, which leads to a collapsed lung

g. <u>Surfactant</u> - A molecule with a hydrophilic end and a hydrophobic end

h. <u>Compliance</u> - The ease with which the lungs inflate

i. <u>Aspirate</u> - To take in by means of suction

j. <u>Tidal volume</u> - The volume of air inhaled or exhaled during normal, quiet breathing

k. <u>Functional residual capacity</u> - The volume of air left in the lungs after a normal exhalation

l. <u>Total lung capacity</u> - The maximum volume of air contained in the lungs after a forceful inhalation

m. <u>Residual volume</u> - The volume of air left in the lungs after a forceful exhalation

2. a. <u>alveoli</u>   b. <u>alveolar duct</u>   c. <u>terminal bronchus</u>   d. <u>uvula</u>   e. <u>diaphragm</u>
f. <u>nasal cavity</u>   g. <u>pharynx</u>   h. <u>larynx</u>   i. <u>trachea</u>   j. <u>primary bronchi</u>

3. The true vocal cords (vocal folds) are inferior to the false vocal cords (the vestibular folds). Thus, air that is leaving the lungs will <u>encounter the vocal folds first</u>.

4. <u>The vestibular folds close off the larynx during deglutition.</u> The epiglottis does this as well. The vestibular folds are simply a "second line of defense."

5. a. <u>The diaphragm and the external intercostals</u> produce normal inspiration, so they are the muscles of principal inspiration.

b. <u>There are no muscles of principal expiration.</u> Normal expiration requires no muscles. It happens when the muscles are relaxed.

c. The sternocleidomastoid, pectoralis minor, and scalene muscles are used to consciously take a deep breath. Thus, they are the muscles of forced inspiration.

d. The abdominal muscles and internal intercostals are used to exhale deeply. Thus, they are the muscles of forced expiration.

6. a. If the person is inhaling forcefully, *both* the muscles of principal inhalation and the muscles of forced inhalation are contracted. Thus, the diaphragm, external intercostals, sternocleidomastoid, pectoralis minor, and scalene muscles are contracted.

b. We inhale by expanding our thoracic cavity. This means the thoracic cavity is large.

c. Since the thoracic cavity is expanded, there is more volume. An increase in volume means a decrease in pressure. Thus, the pressure inside the lungs is lower than the pressure of the atmosphere.

7. The elastic nature of the lungs and the surface tension of the alveolar fluid both aid in the collapse of the lungs. Emphysema can result from a lack of elasticity in the lungs.

8. Negative pressure in the pleural cavity and surfactant in the alveolar fluid keep the lungs from collapsing completely and aid in respiration. A collapsed lung is associated with a loss of negative pressure in the pleural cavity (a pneumothorax), and respiratory distress syndrome is associated with a lack of surfactant in the alveolar fluid.

9. Compliance is lowest at birth. Once the baby begins to breathe for a while, the lungs get stretched, and breathing becomes much easier.

10. External respiration is efficient because of these factors: thin respiratory membrane, large surface area of the alveolus, narrow capillaries, large surface area of the erythrocytes, controlled relationship between ventilation and blood flow through the lungs, and the large functional residual capacity of the lungs.

11. The respiratory membrane includes the following layers, starting at the alveolus and moving to the capillary: the alveolar fluid, the simple squamous epithelium of the alveolus, the basement membrane of the alveolar epithelium, the interstitial space, the basement membrane of the capillary endothelium, and the simple squamous endothelium of the capillary.

12. Pneumonia is a general term that refers to an infection of the lungs. It causes a fluid buildup in the alveoli which reduces the efficiency of external respiration.

13. a. This is a high partial pressure of $CO_2$. Thus, the blood has already received carbon dioxide from the cells. This blood must be heading back to the heart and is therefore in the inferior vena cava.

b. There is only one place that the partial pressure of oxygen in the blood is this high: right before the blood exits the lungs.

c. This is the normal partial pressure of oxygen for blood that has left the lungs (and therefore mixed with blood from the bronchial veins) but has not yet dropped off any oxygen to the tissues. Of the three places mentioned, this blood can only be found in the pulmonary vein between the lungs and the heart.

14. The minimum average partial pressure of oxygen in the blood is about 40 mmHg in resting tissue. Thus, this blood has not given up all of its oxygen yet. However, it has given up a lot, because once it leaves the lungs, the partial pressure of oxygen in blood is 95 mmHg. Thus, the blood must be in the process of giving up oxygen to the tissues. Since that is happening, the blood is also picking up carbon dioxide from the tissues. Its lowest partial pressure is 40 mmHg and its highest is 45 mmHg. Thus, 43 mmHg is in the right range.

15. The Hering-Breuer reflex prevents over-inflation of the lungs by sending inhibitory signals to the respiratory control centers in the medulla as the bronchioles expand. If the inhibitory signals reach the right frequency, the respiratory control centers will stop inspiration.

16. There are respiratory control centers in the medulla and the pons that govern ventilation.

17. The student is wrong because the main control of depth and rate of ventilation comes from the body's monitoring of pH, which tells it the level of carbon dioxide in the body. Oxygen levels affect ventilation only when oxygen drops significantly.

18. If the pH is on the rise, that means the blood is getting to be more basic. Thus, acid must be added. The body can produce more acid in the blood by increasing the amount of carbon dioxide in the blood. This will happen if the rate and depth of ventilation are reduced.

19.

| Step Name | Takes Place In | Makes | |
|---|---|---|---|
| Glycolysis | Cytoplasm | 2 ATP | |
| Oxidation of pyruvate | Outer membrane of mitochondrion | | 2 $CO_2$ |
| Kreb's cycle | Matrix of mitochondrion | 2 ATP | 4 $CO_2$ |
| Electron transport system | Inner membrane of mitochondrion | 32 ATP | 6 $H_2O$ |

There are a total of 36 ATPs made for every glucose molecule in aerobic respiration.

20. a. 2 ADP  b. 2 NADH  c. 4 ATP  d. 2 pyruvate  e. 2 $CO_2$  f. 2NADH
g. 2 acetyl coenzyme-A  h. oxaloacetic acid  i. 6 NAD+  j. 2FADH$_2$  k. 2 ATP  l. 4 $CO_2$
m. 2 $H^+$  n. $H^+$  o. 2$e^-$  p. $e^-$  q. 32 ATP  r. 2 $H^+$  s. $H_2O$

## SOLUTIONS TO THE STUDY GUIDE FOR MODULE #15

1. a. <u>Retroperitoneal</u> – Behind the parietal peritoneum

b. <u>Erythropoiesis</u> – The production of red blood cells (erythrocytes)

c. <u>Renal blood flow rate</u> - The rate at which blood flows through the kidneys (1 liter/min)

d. <u>Filtrate</u> - Blood plasma without proteins, found in the nephrons of the kidneys

e. <u>Glomerular filtration rate</u> - The rate at which filtrate is produced in glomerular filtration (125 mL/minute)

f. <u>Tubular maximum</u> - The maximum rate of reabsorption by active transport through the nephron tubules

g. <u>Buffer system</u> - A mixture of an acid and a base which resists changes in pH

2. a. <u>kidneys</u>  b. <u>ureters</u>  c. <u>urinary bladder</u>  d. <u>urethra</u>  e. <u>renal artery</u>  f. <u>renal vein</u>
g. <u>renal pelvis</u>  h. <u>renal pyramid</u>  i. <u>medulla</u>  j. <u>cortex</u>  k. <u>renal column</u>
l. <u>renal capsule</u>  m. <u>minor calyx</u>

3. The <u>urinary bladder</u> contains stratified transitional epithelium, so that it can stretch.

4. The <u>medulla</u> has the higher concentration of solutes in its interstitial fluid. The medulla's interstitial fluid is as much as <u>four times</u> more concentrated than normal interstitial fluid.

5. The seven functions of the urinary system are: <u>urine formation, pH control, blood pressure regulation, stimulation of red blood cell formation by the red bone marrow, vitamin D activation, transport of urine, and storage and release of urine.</u>

6. <u>When cleaning a desk drawer, you can dump everything out of the desk drawer and then start putting back the things you want to keep. Anything left in the pile at the end is waste. If you make a mistake and put something in the drawer, you can always take it out again and return it to the waste pile. This is like urine formation. The kidney receives all contents of the blood plasma (except proteins), and then it puts back the things that the blood needs. Anything else is waste. If something gets back into the blood, it can be put back into the urine before the process is finished.</u>

7. a. <u>renal corpuscle</u>  b. <u>Bowman's capsule</u>  c. <u>glomerulus</u>  d. <u>descending limb of the loop of Henle</u>  e. <u>loop of Henle</u>  f. <u>ascending limb of the loop of Henle</u>
g. <u>proximal tubule</u>  h. <u>distal tubule</u>  i. <u>cortex</u>  j. <u>medulla</u>  k. <u>collecting duct</u>

8. The proximal tubule and descending limbs of the loop of Henle are always permeable to water. The ascending limb of the loop of Henle is never permeable to water. The distal tubule and collecting duct are permeable to water based on the amount of ADH present.

9. The four steps are filtration, reabsorption, secretion, and water reabsorption. Water reabsorption is often grouped together with reabsorption so that there are only 3 steps to urine formation.

10. In glomerular filtration, filtrate makes it through the filter. Blood cells and proteins do not.

11 A high GFR comes from the high permeability of the glomerular capillaries and the high glomerular capillary pressure.

12. Glomerular capillary pressure is high because the efferent arteriole is thinner than the afferent arteriole. GCP is fought by capsular pressure and colloid osmotic pressure. The difference between GCP and the sum of these two pressures is 7 mmHg. If the GCP loses 7 mmHg of pressure, no more glomerular filtration occurs, which leads to renal shutdown.

13. To be actively reabsorbed, a substance usually needs a carrier and ATP. The exception to this is protein, which needs only ATP, because proteins are absorbed by pinocytosis.

14. Water is the main substance which is passively reabsorbed. Urea is another prime example.

15. If the reabsorption T-max is high, it means that a lot of the substance is actively reabsorbed. As a result, it will go into the blood and not be left in the urine. Thus, you expect to find only a little of it in the urine.

16. If a substance is secreted by the nephron, it is going back into the urine. Thus, the concentration decreases in the blood.

17. a. In the proximal tubule, the filtrate has just entered the nephron. Thus, it has roughly the same concentration of solutes as does blood plasma.

b. As the filtrate descends into the medulla, water leaves the nephron, concentrating the filtrate. Thus, the filtrate has a higher concentration of solutes in the lower portion of the descending limb of the loop of Henle.

c. As the filtrate descends into the medulla, water leaves the nephron, concentrating the filtrate. Thus, the filtrate has a higher concentration of solutes at the bottom of the loop of Henle.

d. As the filtrate rises up the ascending limb of the loop of Henle, solutes are actively transported out of the filtrate, but water is not allowed to follow. Thus, the concentration of solutes in the filtrate is lower than that of plasma.

e. At the distal tubule, the active transport of solute out of the nephron has been going on for quite a while. Thus, the concentration of solutes in the filtrate is lower than that of blood plasma. This can vary, depending on the level of ADH present.

Notice that we did not ask you about the filtrate in the collecting duct. That's because the concentration of solutes in the filtrate at the collecting duct depends on the amount of ADH present.

18. The internal urinary sphincter is controlled automatically. This provides for urination in babies. As we develop, we learn to control the external urinary sphincter, which allows us to decide when we urinate.

19. The juxtaglomerular cells detect and respond to changes in blood pressure and sodium level in the blood.

20. Aldosterone is the hormone stimulated by the secretions of the juxtaglomerular cells. The answer is not renin, angiotensinogen, angiotensin I, or angiotensin II, because these molecules are not hormones. However, angiotensin II stimulates the release of aldosterone from the adrenal cortex, so aldosterone is the *hormone* affected by the juxtaglomerular cells.

21. The atrial natriuretic hormone reduces blood pressure and the sodium level in the blood. It is secreted in the heart atria in response to atrial stretching.

22. When blood pH drops below 7.35, we have acidosis. When blood pH rises above 7.45, we have alkalosis.

23. When a base enters a buffer, the acid in the buffer will react with it. Thus, in the bicarbonate buffer, carbonic acid will react. In the phosphate buffer, dihydrogen phosphate will react.

24. The three processes, in order of effectiveness, are: buffer systems, ventilation depth and speed, and $H^+$ secretion in the kidney.

25. The less effective processes are the faster ones. Thus, in order of increasing speed, we have $H^+$ secretion in the kidney, ventilation depth and speed, and buffer systems.

# SOLUTIONS TO THE STUDY GUIDE FOR MODULE #16

1. a. <u>Spermatogenesis</u> – The process by which sperm form in the testes

b. <u>Erection</u> – The enlarged, firm state of the penis which results as the erectile tissues fill with blood

c. <u>Coitus</u> – Sexual intercourse (the process in which the erect penis enters the vagina)

d. <u>Emission</u> – Movement of the male reproductive secretions towards the urethra

e. <u>Ejaculation</u> – The movement of sperm out of the urethra

f. <u>Semen</u> – A milky-white mixture of sperm and the secretions of the testes, seminal vesicles, prostate gland, and bulbourethral glands

g. <u>Secondary sex characteristics</u> – The characteristics which appear at puberty and tend to distinguish men from women. These include the development of breasts, hairline patterns, facial shape, body shape, and the distribution of body hair.

h. <u>Puberty</u> – A series of events which transforms a child into a sexually mature adult

i. <u>Anabolism</u> – All of the synthesis reactions which occur in the body

j. <u>Catabolism</u> – All of the decomposition reactions which occur in the body

k. <u>Oogenesis</u> – The production of haploid germ cells by the ovary

l. <u>Ovulation</u> – The release of a secondary oocyte from a mature follicle

m. <u>Menopause</u> - The last menstruation; the time of life after which a woman no longer ovulates

n. <u>Lactation</u> - The process by which a female mammal produces and secretes milk to feed her young

o. <u>Menses</u> - Periodic shedding of the uterine endometrium which results in blood loss from the vagina

p. <u>Implantation</u> - Attachment of the blastocyst to the endometrium

q. <u>Organogenesis</u> - The formation of organs in a developing fetus

2. a. <u>vas deferens</u>  b. <u>erectile tissue</u>  c. <u>penis</u>  d. <u>glans penis</u>  e. <u>foreskin or prepuce</u>
f. <u>seminal vesicle</u>  g. <u>prostate gland</u>  h. <u>bulbourethral gland</u>  i. <u>urethra</u>  j. <u>epididymis</u>
k. <u>testis</u>  l. <u>ovary</u>  m. <u>uterus</u>  n. <u>cervix</u>  o. <u>clitoris</u>  p. <u>vagina</u>

3. The proper order is: <u>testis, epididymis, vas deferens, male urethra, vagina, uterus, and uterine tube</u>.

4. Body temperature is too warm for spermatogenesis. Thus, <u>the testes must be in the scrotum to stay at a lower temperature than body temperature</u>.

5. a. Spermatogonia are diploid cells. Thus, they have <u>46 chromosomes</u>.

b. Primary spermatocytes are the product of mitosis. Thus, they have <u>46 chromosomes</u>.

c. Secondary spermatocytes are the result of meiosis I. Thus, they have <u>23 chromosomes</u>.

d. Spermatids are the result of meiosis II. Thus, they have <u>23 chromosomes</u>.

e. Sperm are just mature spermatids. Thus, they have <u>23 chromosomes</u>.

f. A primary oocyte is stuck in prophase I of meiosis. Thus, it has <u>46 chromosomes</u>.

g. A secondary oocyte has completed meiosis I. Thus, it has <u>23 chromosomes</u>, though they have a duplicate for each.

h. A zygote is the union of a sperm and egg. Thus, it has <u>46 chromosomes</u>.

6. <u>Sertoli cells support and nourish the cells performing spermatogenesis, and Leydig cells produce and secrete testosterone</u>.

7. <u>In males, FSH mostly affects Sertoli cells and LH mostly affects Leydig cells</u>.

8. <u>There are four sperm from every one primary spermatocyte. You can only form one egg from one oogonium, however</u>.

9. <u>Males have no theoretical limit to the number of sperm they can produce. Women, however, are limited by the number of primary oocytes with which they are born</u>.

10. FSH and LH are both on the rise only during the end of <u>the follicular stage</u>. Since the follicles are still being stimulated, <u>ovulation has not occurred</u>.

11. Progesterone stimulates the thickening and development of the endometrium. Thus, <u>the endometrium is thickening</u>. Since progesterone is on the rise and estrogen is decreasing, this is <u>the luteal stage</u>.

12. If fertilization does not occur, <u>the endometrium is sloughed off, and the woman menstruates. This occurs because the corpus luteum degenerates, and its secretion of estrogen and progesterone drops</u>.

13. Fraternal twins are the result of two eggs being fertilized during the same cycle. This can happen if the woman ovulates more than one secondary oocyte, which is not common. Identical twins form during cleavage, when the cells separate and begin forming two different embryos.

14. The sex of the child is determined by whether or not there is a Y chromosome. Eggs can have only X chromosomes, but sperm can have the X or the Y chromosome. If a sperm with the Y chromosome fertilizes the egg, the child is male. If a sperm with an X chromosome fertilizes the egg, the child is female.

15. a. This is a single diploid cell that results from fertilization.

b. In this stage, the zygote begins reproducing by mitosis, but the size of the embryo does not increase.

c. In this stage, the cells of the embryo form a roughly spherical shape. This usually happens with about 12 cells.

d. In this stage, a fluid-filled cavity forms in the middle of the embryo cells. The cells become either trophoblasts or part of the inner cell mass.

e. In this stage, the embryo changes shape and the cells differentiate into the ectoderm, mesoderm, and endoderm.

f. In this stage, the nervous system begins to form.

16. Implantation occurs in the blastula phase.

17. The placenta forms from the trophoblasts.

# Tests

# TEST FOR MODULE #1

1. Define the following terms:

   a. Physiology
   b. Histology
   c. Organ
   d. Selective permeability
   e. Endocytosis

2. Of the seven levels of organization we discussed, which would fall under a study of gross anatomy?

3. Identify the type of tissue that makes up the following:

   a. The spinal cord
   b. The lining of the larynx
   c. The biceps muscle

4. The level of $Ca^{2+}$ ions in the blood must be carefully regulated. Cells in the parathyroid gland monitor the $Ca^{2+}$ level. If the $Ca^{2+}$ level decreases, the parathyroid glands are stimulated to start producing more PTH, a hormone. When PTH hormone levels are elevated, the kidneys will increase the amount of $Ca^{2+}$ returned to the blood during the urine-making process. This results in an increase of $Ca^{2+}$ in the blood.

   a. Is this positive or negative feedback?
   b. What is the control center?
   c. What is the effector?

5. What organelle makes most of the ATP in the cell?

6. List the phases of mitosis in order.

7. How do small polar molecules, like water, enter a cell?

8. How do larger molecules, like carbohydrates, enter a cell?

9. How do very large molecules, like proteins, enter a cell?

10. A sodium ion travels from inside the cell to outside the cell. The cell uses ATP to make this happen. Where is the concentration of sodium ions greater: inside or outside the cell?

11. Is endocytosis an example of active or passive transport?

# TEST FOR MODULE #2

1. Define the following terms:

   a. Endocrine glands
   b. Fibroblasts
   c. Chondrocytes
   d. Stromal cells
   e. Labile cells

2. Identify the epithelial tissue below. Give its function and a place it can be found in the body.

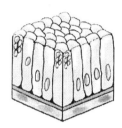   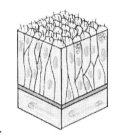

    a.                     b.

3. Where is the free surface in #2a? Where is the basement membrane?

4. An exocrine gland secretes a substance that is rich in cytoplasm and phospholipids but no other cellular material. What kind of exocrine gland is it?

5. Looking at some tissue under a microscope, you notice that the cells are in little pockets. Is this connective tissue proper or cartilage?

6. Identify the connective tissue proper by the sketch of its microscopic structure. Give its function and one place in the body that it is found.

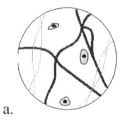   

    a.                     b.

7. A membrane is found in a joint. What kind of membrane is it? What is its function?

8. An organ's parenchymal cells are stable. If the organ gets damaged, can it be repaired to 100% working order again? Is the healing process slow or relatively quick?

9. How do epithelial cells get oxygen and nutrients?

10. Identify the cartilage by the sketch of its microscopic structure. Give its function and one place in the body that it is found.

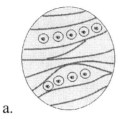

a.

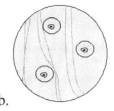

b.

# TEST FOR MODULE #3

1. Define the following terms:

   a. Hemopoiesis
   b. Cancellous bone
   c. Axial skeleton
   d. Suture
   e. Process

2. What layer of skin is superficial to the dermal papillae?

3. An epidermal skin cell is keratinized. What layers of the epidermis is it NOT in?

4. If a hair cell contains *no* keratin, where is it?

5. Identify the structures pointed out in the skin cross section below:

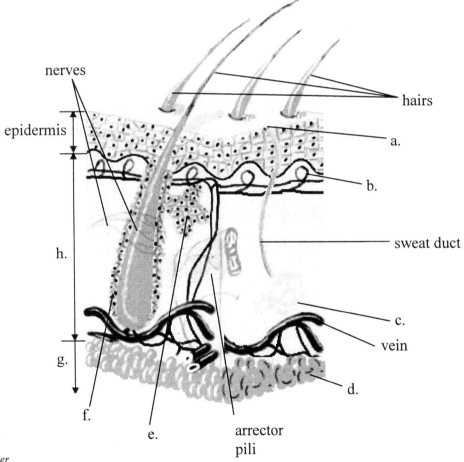

*Illus. by Megan Whitaker*

66 The Human Body: Fearfully and Wonderfully Made

6. Label the bones in the figure below:

*Illus. by Megan Whitaker*

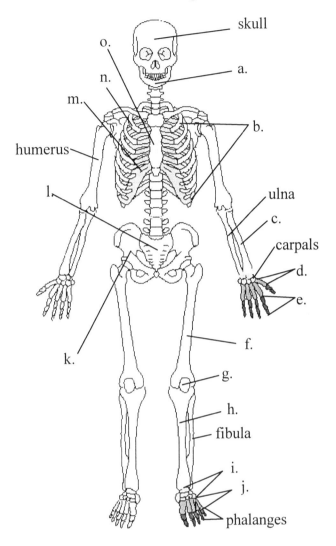
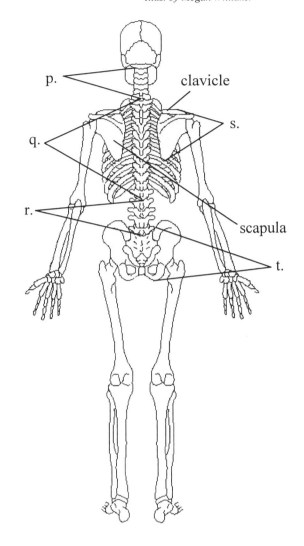

7. Label the bones in the figure below:

*Illus. by Megan Whitaker*

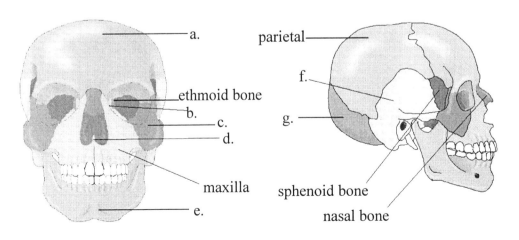

# TEST FOR MODULE #4

1. Define the following terms:

a. Osteoblast
b. Osteocyte
c. Osteoclast
d. Hematoma
e. Callus
f. Anatomical position

2. A bone is low on collagen. What will that do to the properties of the bone?

3. What kind of bone tissue contains trabeculae?

4. By means of what structure do osteocytes communicate with one another?

5. Name three reasons that bone must be continually remodeled.

6. What kind of growth is available to a bone whose epiphyseal plate is ossified? You should use the proper term.

7. When a bone is being repaired, what holds the pieces of bone together while the repair occurs?

8. A person's body starts actively secreting PTH. What gland is working, and why is this happening?

9. What hormones cause a rapid growth spurt in a child followed by no more growth?

10. What two types of joints in the body provide for little or no motion?

11. Which type of synovial joint offers the greatest range of motion?

12. Which type of synovial joint offers the smallest range of motion?

13. Name the kind of motion exhibited in the following actions:

a. A person is doing squats. He is standing up straight and then bends his knees so that his body is lowered.
b. A person is standing on the ground and then bends his ankle so that his toes are pointing upwards.
c. A person twirls his arms in a circular motion.

14. Label the parts of the following illustration.

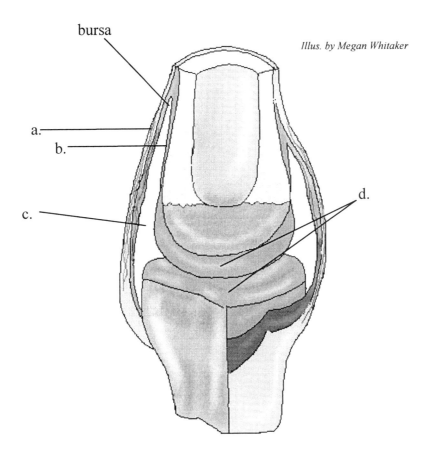

# TEST FOR MODULE #5

1. Define the following:

a. Sarcomere
b. Motor unit
c. All-or-none law of skeletal muscle contraction
d. Muscle tone

2. Muscle tissue is striated when examined under a microscope. What kind of muscle tissue is this?

3. Label the following structures of a muscle:

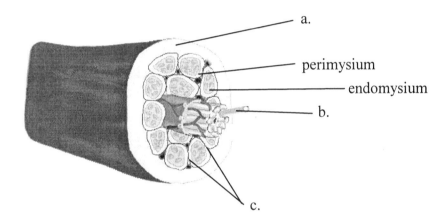

4. Label the parts of the sarcomere below:

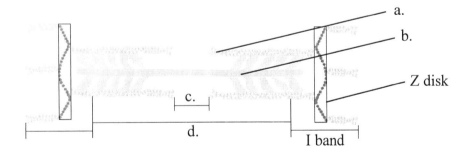

5. Using the figure above, indicate which things remain the same when a sarcomere contracts.

a. The length of the structure labeled (a) in the figure
b. The length of the structure labeled (b) in the figure
c. The length of the structure labeled (c) in the figure
d. The length of the structure labeled (d) in the figure
e. The distance between the Z disks

6. Fill in the blanks

Without _____ to destroy _____, a muscle cell would stay contracted.

7. How does the membrane action potential get to the sarcoplasmic reticulum?

8. Calcium ions are diffusing across the membrane of the sarcoplasmic reticulum. Is the cell beginning contraction or ending contraction?

9. When a sarcomere contracts, what happens between the power stroke and the return stroke?

10. What must happen after the action potential reaches the presynaptic terminal and before an action potential is established on the membrane of a muscle cell?

11. Two muscle fibers lie side-by-side but contract at different times. Are they a part of the same motor unit?

12. Fill in the blank:

If a muscle is stiff, there is not enough _____ in the muscle cells.

13. A nerve is being stimulated with a submaximal stimulus. Are there any motor units that have been recruited?

14. A motor unit is recruited. What happens to the muscle fibers in that motor unit?

15. Lactic acid is building up in a muscle cell. What kind of respiration is occurring?

16. A muscle cell is using creatine phosphate to convert ADP into ATP. Does this require oxygen?

# TEST FOR MODULE #6

1. Define the following terms:

a. Origin
b. Insertion
c. Belly
d. Mastication

2. Identify the muscles in the following figure:

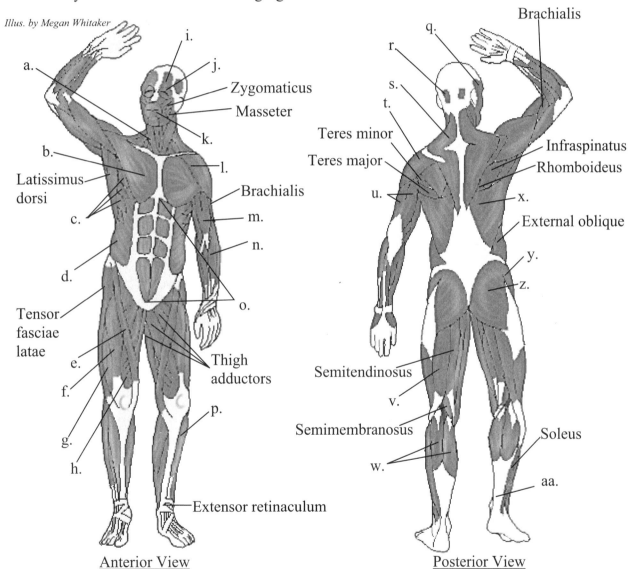

3. The orbicularis oris and what other muscle are called the kissing muscles?

4. Of the muscles in the figure above, name one that acts on the vertebral column.

5. Of the muscles in the figure for question 2, name one that flexes the forearm. Name the antagonist of this muscle.

6. Which of the structures above are not muscles?

7. Name one of the muscles in the figure above that flexes the thigh. Name an antagonist of this muscle.

# TEST FOR MODULE #7

1. Define the following terms:

a. Ganglia
b. Spinal nerves
c. Efferent neurons
d. Somatic motor nervous system
e. Association neuron
f. Excitability

2. Identify the parts of the neuron below:

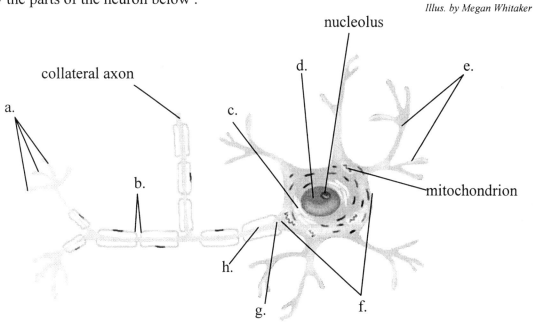

*Illus. by Megan Whitaker*

3. How many axons does a multipolar neuron have?

4. What is the purpose of a non-ciliated ependymal cell?

5. A nerve sends a signal from the CNS to the smooth muscles of the intestine. Is this a part of the autonomic nervous system or the somatic nervous system?

6. What two conditions must be met for an axon to regenerate when severed? Do axons in the CNS regenerate when severed?

7. During depolarization in an axon, what is happening to the sodium ions? What is happening to the potassium ions?

8. The concentration of potassium ions outside an axon is very high, and the concentration of sodium ions inside is very high. Is the potential difference between the inside and outside of the axon negative or positive?

9. What has to happen to the axon in problem #8 in order to get back to the resting potential?

10. A stimulus on a neuron does not result in an action potential. There are two reasons why this might not happen. What are those reasons?

11. An action potential travels along an axon by skipping from node of Ranvier to node of Ranvier. What is this kind of conduction called? Is this faster or slower than an action potential running down an axon by continuous conduction?

12. You are listening to the radio. Suddenly, your little brother turns the radio up really loud, so that the noise actually hurts your ears. Compare the maximum potential difference of *each individual action potential* before and after your little brother turned the volume up.

13. In the situation described in problem #12, compare the frequency of action potentials running from your ears to your CNS before and after your little brother turned the volume up.

14. A signal is stimulated at a receptor and travels to the CNS. The number and frequency of action potentials is exactly the same at the CNS as it was at the receptor. Did this signal travel through a synapse?

15. A signal travels down an axon and then encounters an excitatory synapse. Compare the frequency of action potentials in the presynaptic neuron and the postsynaptic neuron.

16. At a synapse, the release of neurotransmitters results in an opening of potassium channels at the postsynaptic membrane. Is this an inhibitory synapse or an excitatory synapse?

17. A signal originates in one receptor and ends up creating action potentials in many different places in the CNS and PNS. What kind of neuron arrangement did it pass through?

# TEST FOR MODULE #8

1. Define the following:

   a. Gray matter
   b. White matter
   c. Decussation
   d. Vital functions
   e. Commissures

2. Identify the structures in the following figure:

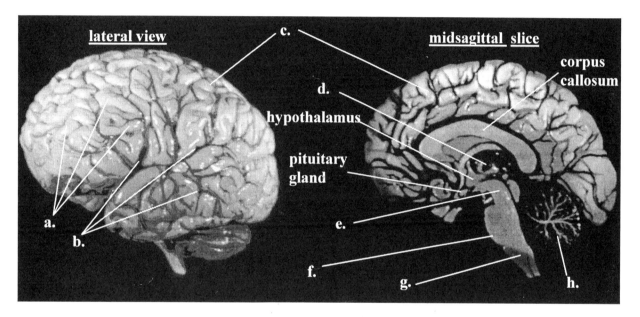

3. What is the structure made by letters e, f, and g in the figure above?

4. What is the structure formed by both d and the hypothalamus in the figure?

5. What is the main function of the ventricles in the brain?

6. Explain the purpose of a reflex arc.

7. Explain how a basic reflex arc works.

8. Explain how the brain can control a skeletal muscle which is also controlled by a reflex arc.

Match the functional areas of the cerebral cortex with their functions:

| Functional Area | Function |
|---|---|
| 9. primary somatic sensory area | a. controls skeletal muscle movements |
| 10. somatic sensory association area | b. deals with the comprehension of speech |
| 11. visual association area | c. interprets the meaning of sound by placing it into context with your past experiences |
| 12. visual cortex | d. interprets the sensory information and puts it into context with your past experiences |
| 13. Wernicke's area | e. interprets the basic visual information such as shape and color |
| 14. auditory association area | f. works out the sequence of signals needed for complex motion |
| 15. primary auditory area | g. initiates the muscle movements for speech |
| 16. Broca's area | h. recognizes the meaning of visual information by putting it into context with your past experiences |
| 17. taste area | i. receives and localizes general sensations from the entire body |
| 18. prefrontal area | j. interprets taste |
| 19. premotor area | k. site of motivation and foresight; regulates mood and emotion |
| 20. primary motor cortex | l. interprets the basics of sound such as pitch and volume |

# TEST FOR MODULE #9

1. Define the following terms:

a. Sensory receptor
b. Thermoreceptors
c. Nociceptors
d. Cutaneous receptors
e. Proprioceptors

2. A nerve pathway leaving the spinal cord is composed of two neurons which synapse at a ganglion. The axon of the first neuron is long, while the axon of the second is short. Is this part of the SMNS or the ANS? In addition, state more specifically to what division of the nervous system it belongs.

3. Name three cutaneous receptors and list their functions.

4. In order for us to smell a substance, it must be volatile, partly water soluble, and partly fat soluble. List the reasons for each of those conditions.

5. Identify the structures in the following figure:

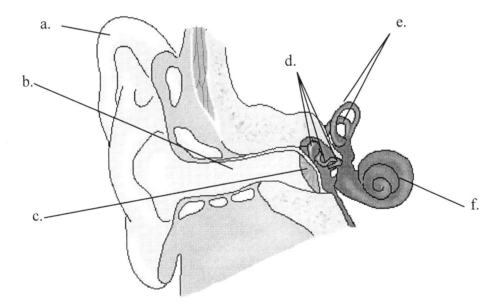

6. Of the structures listed above, which contains the cupula?

7. Which structure in the figure above contains the organ of hearing?

8. When a sound wave hits your ear, which is the first structure that begins to vibrate? Which is the last structure?

9. Identify the structures in the following figure, and list the function of each.

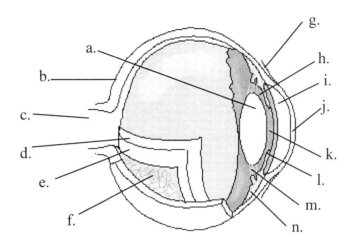

10. Which of the structures above are involved in the process of accommodation?

11. Which two structures above are most important in terms of bending light for the purpose of focusing?

12. In the picture above, where are the rods and cones found? Although not pictured above, where are the cones concentrated?

13. A person has trouble maintaining balance while standing still. Of the structures listed below, which should be investigated as the source of the problem?

   the crista ampullaris, the ampullae, the ultricular macula, the external auditory meatus

# TEST FOR MODULE #10

1. Define the following:

   a. Neurosecretory cells
   b. Prostaglandins

2. Identify the glands in the following figure:

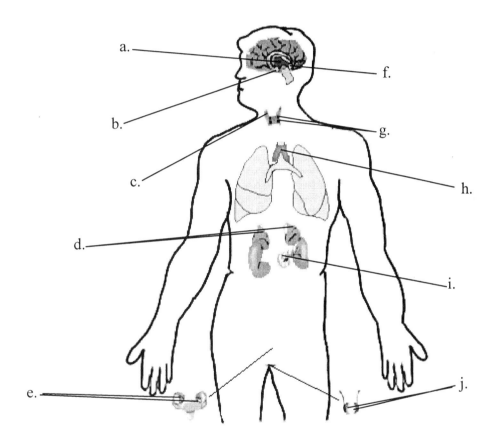

3. The following graph reports the level of hormone in the body as a function of time. What secretory pattern is this?

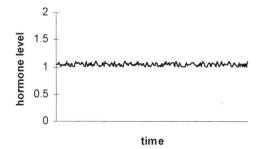

Match the hormone to its function. Also, name the gland from which it is secreted:

4. Growth hormone releasing hormone (GH-RH)  i
5. Parathyroid hormone (PTH)  k
6. Cortisol  m
7. Testosterone  j
8. Thyroid stimulating hormone (TSH)  e
9. Prolactin (PRL)  g
10. Melatonin  f
11. Epinephrine (E)  l
12. Oxytocin (OT)  a
13. Gonadotropin releasing hormone (GnRH)  b
14. Prolactin inhibiting hormone (PIH)  c
15. Antidiuretic hormone (ADH)  h
16. Growth hormone (GH)  d
17. Glucagon  p
18. Estrogen  q
19. Aldosterone  n
20. Insulin  o

a. Increases the contractions of the uterus during birth and promotes the release of breast milk.
b. Increases the release of FSH and LH from the anterior pituitary
c. Decreases the release of PRL from the anterior pituitary
d. Increases growth in most tissues
e. Increases the release of thyroxin from the thyroid gland
f. Affects release of GnRH by hypothalamus
g. Stimulates milk production in the breasts
h. Increases the retention of water by the kidneys
i. Increases the release of GH from the anterior pituitary
j. Sex hormone in males
k. Increases blood calcium levels by increasing osteoclast activity
l. Enhances sympathetic response
m. Increases protein and fat breakdown in most tissues
n. Increases the retention of sodium by the kidneys
o. Lowers blood glucose by stimulating cells to absorb glucose
p. Raises blood glucose by causing liver to release glucose
q. Sex hormone in females

21. A hormone stimulates a cell to absorb chemicals which are outside of the cell. Did this hormone stimulate a membrane-bound receptor or an intracellular receptor?

# TEST FOR MODULE #11

1. Define the following terms:

a. Erythrocytes
b. Leukocytes
c. Platelets
d. Hemopoiesis
e. Systolic phase
f. Diastolic phase
g. Arterioles
h. Venules

2. Which of the following is NOT true about blood?

a. More dense than water
b. pH ranges between 6.5 and 7.5
c. Composed of both water and cells
d. Carries regulatory chemicals, ions, gases, proteins, and wastes

Match the following with its description:

3. platelet
4. basophil
5. erythrocyte
6. eosinophil
7. monocyte
8. neutrophil
9. lymphocyte

a. Granulocyte that fights infections by phagocytosis
b. Agranulocyte that fights infections by phagocytosis
c. Cell fragment in blood which helps prevent blood loss
d. Cell that carries oxygen in the blood
e. Cell that produces antibodies
f. Cell that releases histamine and heparin
g. Cell that fights inflammation

10. A region in a blood vessel has prothrombinase in it. Which stage of hemostasis is taking place?

11. Of the following chemicals, which exist *only* when the blood coagulation process is occurring?

factor IX, activated factor XII, fibrinogen, calcium ions, thrombin, fibrin

12. In what type of blood will you find *no antibodies* against erythrocyte or Rh antigens?

13. In what type of blood could you find antibodies against the B antigen and the Rh antigen, but no antibodies against the A antigen?

14. A father with type AB+ blood has a child with a mother whose blood type is A-. The child has a blood type of B-. What blood type alleles and Rh factor alleles does the father have? What blood type alleles and Rh factor alleles does the mother have?

15. Identify the structures in the following anterior view of the heart:

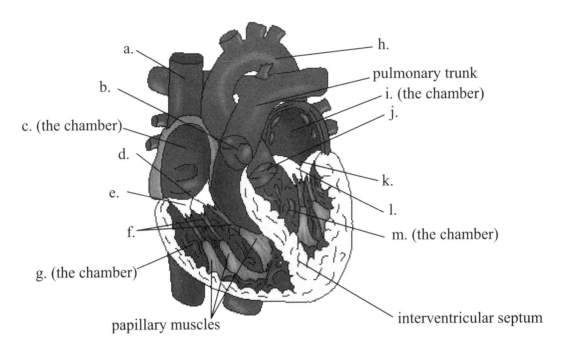

16. During systole, indicate whether the atria are contracted or relaxed, whether the ventricles are contracted or relaxed, and whether each valve is open or closed.

17. Fill in the blanks with structures from the diagram in #15 or the words oxygenated or deoxygenated:

Blood that travels through the aorta is _____. It travels to the tissues of the body and returns to the heart. When it reaches the heart, it is _____. It is dumped into the _____. It then goes through the _____ and _____ so that it can go into the _____. Then, it passes through the _____ and into the pulmonary trunk so that it can travel to the lungs. In the lungs, the blood becomes _____. It comes back from the lungs and is dumped into the _____. It then travels through the _____ and _____ to get to the _____. Finally, it passes through the _____ to get back into the aorta.

# TEST FOR MODULE #12

1. Define the following terms:

   a. Lymph nodes
   b. Diffuse lymphatic tissue
   c. Innate immunity
   d. Acquired immunity
   e. Humoral immunity
   f. Cell-mediated immunity

Match each of the following to the letter which best describes it

2. interstitial fluid
3. lymph
4. tonsils
5. Peyer's patches
6. lymph node
7. spleen
8. thymus gland
9. vasodilation
10. pyrogens
11. interferon
12. complement
13. constant region
14. variable region
15. memory B-cell
16. plasma B-cell
17. cytotoxic T-cell
18. helper T-cell
19. memory T-cell
20. MHC

a. Cleans the lymph as it travels back to the bloodstream
b. Chemicals that affect the hypothalamus, increasing body temperature
c. An innate immune response, causing cells to increase antiviral defenses
d. Fluid that flows in lymph vessels
e. Section of the antibody that determines what group (IgG, IgA, etc.) it is in
f. Responsible for the secondary response of the humoral immune system
g. Increases proliferation of T-cells and B-cells
h. Groups of lymph nodules found on the small intestine
i. A group of glycoproteins that identifies cells as a part of the body
j. Responsible for the secondary response of the cell-mediated immune system
k. Fluid that exists between the cells
l. Attacks and lyses foreign cells as a part of cell-mediated immunity
m. Produces antibodies to fight an infection currently in the body
n. Place where T-lymphocytes mature
o. Groups of lymph nodules found in the throat and on the back of the tongue
p. Section of the antibody that determines what antigen it can bind to
q. An antibacterial response based on twenty proteins in the blood serum
r. A result of inflammation which causes increased blood flow
s. Cleans the blood

# TEST FOR MODULE #13

1. Define the following terms:

a. Deglutition
b. Peristalsis
c. Lumen
d. Micronutrients

2. Label the organs in the following figure:

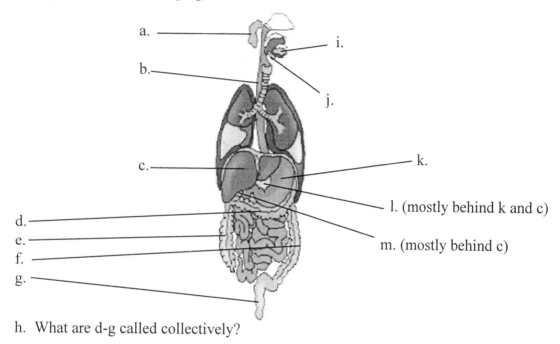

h. What are d-g called collectively?

3. Name the structure that is best described by the following:

a. Stores excess glucose as glycogen
b. The first source of amylase secreted into food
c. The second source of amylase secreted into food
d. Produces bile
e. Has gastric pits that secrete gastric juice
f. Holds feces and triggers the defecation reflex
g. Most nutrient absorption takes place here
h. Has endocrine and exocrine functions and is an accessory organ
i. Contains glands which produce CCK
j. Contains glands which produce secretin
k. Holds bile and concentrates it
l. Moves food from the pharynx to the stomach
m. Interconverts macronutrients based on the body's needs

4. The epiglottis and soft palate both participate in deglutition. What do they do?

5. What are intestinal villi? Why are they so important?

6. What is wrong with the following statement?

"Certain cells in the stomach produce pepsin, an enzyme that breaks down proteins."

7. Without intrinsic factor, what can the body not do?

8. What are the three classes of macronutrients?

9. What are the two vitamins which can be produced in the body as well as absorbed from food?

# TEST FOR MODULE #14

1. Define the following terms:

   a. Lower respiratory tract
   b. Ventilation
   c. Internal respiration
   d. Pneumothorax
   e. Compliance
   f. Tidal volume
   g. Functional residual capacity

2. Identify the organs in the following figure:

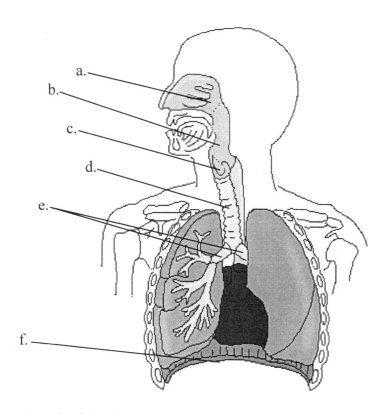

Match each of the following respiratory abnormalities with its description

3. Pneumonia
4. Respiratory distress syndrome
5. Emphysema
6. Pneumothorax

   a. Difficulty exhaling due to lack of elastic tissue
   b. A loss of negative pressure in the pleural cavity
   c. A lung infection which builds up fluid in the alveoli
   d. Difficulty inhaling due to a lack of surfactant in the alveolar fluid

7. A person's diaphragm is contracted. What is the pressure in the lungs compared to that of the atmosphere while the air is still flowing?

8. A person's abdominal muscles are tightly contracted for breathing purposes. Is the person inhaling or exhaling? Is the breathing normal or forced?

9. A person's sternocleidomastoid and scalene muscles are contracted as far as possible for the purpose of breathing. Is the volume of air in the lungs closest to the tidal volume, residual capacity, functional residual capacity, or total lung capacity?

10. A person's airway becomes partially blocked due to an obstruction. Of the six factors which increase the efficiency of external respiration, which is affected?

11. In going from the lumen of the alveolus to the lumen of a capillary, a molecule of oxygen encounters the alveolar fluid, then the epithelium of the alveolus, then the interstitial space, then the endothelium of the capillary. What two layers of the respiratory membrane were left out of this description?

12. In the Hering-Breuer reflex, receptors in the bronchioles send signals to what part of the brain? Are the signals excitatory or inhibitory?

13. Blood pH is falling. What can the respiratory system do to help rectify the situation?

14. Fill in the blanks:

The first stage of aerobic respiration is (a)_____. In this stage, glucose is broken down into two molecules of (b)_____, which then enter the second stage of aerobic respiration. It takes two molecules of (c)_____ to start the process, but by the end, (d)___ molecules of that same substance are made. In addition, 2 molecules of NAD+ are converted to 2 molecules of (e)_____, which head to the final stage of aerobic respiration.

The second stage of aerobic respiration is (f)_____. In this stage, the products of the first stage are converted to acetyl coenzyme-A. This results in two molecules of (g)_____ being converted to NADH, which head to the final stage of aerobic respiration.

The third stage of aerobic respiration is the (h)_____. In this stage, the acetyl coenzyme-A is reacted with (i)_____, which, after a series of reactions, is produced again, ready to start the next cycle. In the process, four molecules of (j)_____ are produced, along with six molecules of NADH and two molecules of (k)_____, which all head to the final stage of aerobic respiration.

The final stage of aerobic respiration is (l)_____. In this stage, the (m)_____ and (n)_____ made in the previous stages release their electrons and their (o)_____. The electrons travel through a chain of carrier proteins that use the energy of the electrons to transport hydrogen ions over the inner mitochondrial membrane. When the hydrogen ions come back through the membrane, 32 molecules of (p)_____ are formed.

# TEST FOR MODULE #15

1. Define the following terms:

   a. Retroperitoneal
   b. Erythropoiesis
   c. Renal blood flow rate
   d. Glomerular filtration rate
   e. Tubular maximum
   f. Buffer system

2. Starting with the proximal tubule, order the structures listed below in terms of when they are encountered by filtrate as it travels out of the body.

   loop of Henle, renal pelvis, distal tubule, proximal tubule, major calyx, collecting duct, minor calyx

3. Label the structures pointed out below:

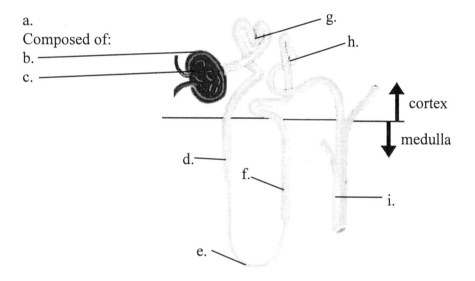

4. Which of the structures above change their permeability to water? What causes the change?

5. In the drawing below, the arrows indicate the direction in which solutes travel between a capillary and the nephron. Indicate whether each arrow illustrates reabsorption or secretion.

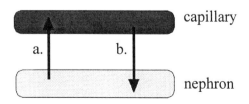

6. An increase in GCP will do what to GFR?

7. Two substances are in the filtrate in equal concentration at the proximal tubule and both exceed their T-max concentration. If the reabsorption T-max is higher for substance A than substance B, compare the concentrations of A and B in the blood as it leaves the kidney.

8. What are juxtaglomerular cells and what do they do in the body?

9. What is the normal range for the pH of blood?

10. The pH of blood is 7.30. Is the person in acidosis or alkalosis?

11. Describe the three processes which regulate blood pH. Indicate the relative effectiveness of each one as well as the relative speed of each one.

12. In the bicarbonate buffer, which substance reacts if an acid is introduced in the blood?

Tests 91

# TEST FOR MODULE #16

1. Define the following terms:

   a. Spermatogenesis
   b. Semen
   c. Secondary sex characteristics
   d. Puberty
   e. Anabolism
   f. Catabolism

2. Identify the structures in the following figures:

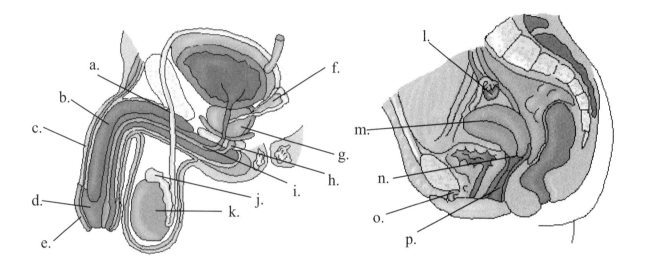

Match each of the following cells to the phrase that describes it best.

3. spermatogonium     a. A cell that nourishes and feeds other cells during spermatogenesis

4. secondary oocyte   b. A cell that will eventually be a part of the placenta in a fetus

5. Sertoli cell       c. One of the three types of cells formed during the gastrula phase

6. mesoderm cell      d. A cell that secretes testosterone

7. Leydig cell        e. A cell that begins spermatogenesis by undergoing mitosis

8. spermatid          f. A cell that will not complete meiosis II unless fertilization occurs

9. trophoblast        g. A cell in the testes that has 23 chromosomes

10. When fertilization occurs, do the egg and cell contribute equal amounts of genetic material? If not, indicate which contributes more.

11. When fertilization occurs, do the egg and cell contribute equal amounts of cytoplasm? If not, indicate which contributes more.

12. The amount of FSH and LH are both decreasing in a woman. Has ovulation occurred?

13. Which ovarian hormone dominates the luteal phase of the menstrual cycle?

14. List the following stages of development in the proper order:

        cleavage, zygote, gastrula, neurula, blastula, morula

15. Which of the stages above are completed while the embryo is outside of the uterus?

# Solutions To The

# Tests

# SOLUTIONS TO THE TEST FOR MODULE #1

1. a. <u>Physiology</u> - The study of the functions of an organism and its parts

b. <u>Histology</u> - The study of tissues

c. <u>Organ</u> - A group of tissues specialized for a particular function

d. <u>Selective permeability</u> - The ability to let certain materials in or out while restricting others

e. <u>Endocytosis</u> - The process by which large molecules are taken into the cell

2. Gross anatomy refers to the macroscopic structures in the body. This would include the <u>organism, organ systems, and organs</u>. Once you hit tissues, you are dealing with things you must see under a microscope, which is microscopic anatomy.

3. a. The spinal cord is made of <u>nervous tissue</u>.

b. Organ lining is typically <u>epithelial tissue</u>.

c. Muscles are made of <u>muscle tissue</u>.

4. a. This is <u>negative feedback</u>, because the result is opposite of the stress.

b. The <u>parathyroid glands</u> are the control center, as it reacts to the stress.

c. The <u>kidneys</u> are the effectors, as they make the blood $Ca^{2+}$ level increase.

5. The <u>mitochondrion</u> is the power plant of the cell. This means it makes energy, and cells store energy as ATP. Thus, the mitochondrion makes most of the cell's ATP.

6. <u>Prophase, metaphase, anaphase, telophase</u>. If the student listed interphase at the beginning, that's fine.

7. Small polar molecules enter the cell through <u>channel proteins</u>.

8. Larger molecules enter the cell through <u>carrier proteins</u>.

9. Very large molecules enter through <u>pinocytosis</u>. Endocytosis deserves half credit, but the student should have been more specific.

10. Since the cell used ATP, this is active transport. This kind of transport is active if it goes against the dictates of diffusion. Thus, the sodium ion had to travel from *low* concentration to *high* concentration. This means the concentration of sodium ions is greater <u>outside of the cell</u>.

11. Endocytosis is *always* active transport.

# SOLUTIONS TO THE TEST FOR MODULE #2

1. a. <u>Endocrine glands</u> - Ductless glands that secrete hormones into the bloodstream

b. <u>Fibroblasts</u> - Spindle-shaped cells that form connective tissue proper

c. <u>Chondrocytes</u> - Mature cartilage cells

d. <u>Stromal cells</u> - Cells that provide structure or support for parenchymal cells

e. <u>Labile cells</u> - Cells that undergo mitosis regularly and quickly


2. a. This is <u>simple columnar epithelium</u>. It can be found in <u>the stomach and intestines</u> (you only needed to mention one of those). Its function is <u>complex absorption and secretion as well as protection</u>.

b. This is <u>pseudostratified epithelium</u>. It can be found in <u>the airways of the lungs</u>. Its function is <u>to produce and move mucus</u>.

3. <u>The free surface is the top of the tissue. The basement membrane is between the cells and the purple tissue beneath the cells.</u>

4. This is an <u>apocrine gland</u>. The phospholipids come from the plasma membrane, and the cytoplasm means there is cellular material in the secretion. However, there is no other cellular material, so only a portion of the cell is in the secretion. That means it is apocrine, not holocrine.

5. This is <u>cartilage</u>. The ground substance of cartilage is firm, requiring that the chondrocytes live in lacuna.

6. a. This is <u>loose connective tissue</u>. It provides for <u>light-duty binding</u>. It can be found <u>under the skin</u>.

b. This is <u>adipose tissue</u>. It provides for <u>insulation and bracing organs</u>. It can be found in the <u>mammary glands, under the skin, and around the kidneys</u> (you need to mention only one).

7. <u>Synovial membranes</u> are found around joints. They provide <u>lubrication for the joint</u>.

8. <u>The organ can be repaired to 100% working order</u>. Since the functional cells are stable, that means they *can* reproduce. However, the <u>healing will be slow</u>, because stable cells do not undergo mitosis regularly.

9. Epithelial cells get oxygen and nutrients through <u>diffusion</u>, because the basement membrane is avascular.

10. a. This is <u>fibrocartilage</u>. It can be found in the <u>joints of the backbone</u>. It provides <u>tough binding and resilient support</u>.

b. This is <u>elastic cartilage</u>. It can be found in the <u>outer ear</u>. It provides <u>flexible support</u>

# SOLUTIONS TO THE TEST FOR MODULE #3

1. a. <u>Hemopoiesis</u> - The process of manufacturing blood cells

b. <u>Cancellous bone</u> - Bone with many small spaces or cavities surrounding the bone matrix

c. <u>Axial skeleton</u> – The portion of the skeleton that supports and protects the head, neck, and trunk

d. <u>Suture</u> - A junction between flat bones of the skull

e. <u>Process</u> - A projection on a bone

2. The <u>epidermis</u> sits on top of the dermal papillae. Thus, it is superficial to them.

3. The two deepest layers, the <u>stratum basale</u> and the <u>stratum spinosum</u>, contain living cells. Thus, a keratinized cell cannot be in those layers.

4. All differentiated hair cells contain keratin. Thus, this must be in the <u>matrix</u>.

5. a. <u>sweat pore</u>  b. <u>blood vessel loop in dermal papillae</u>  c. <u>sweat gland</u>  d. <u>adipose tissue</u>  e. <u>sebaceous gland</u>  f. <u>hair follicle</u>  g. <u>hypodermis</u>  h. <u>dermis</u>

6. a. <u>mandible</u> b. <u>thoracic cage</u> c. <u>radius</u> d. <u>metacarpals</u> e. <u>phalanges</u> f. <u>femur</u> g. <u>patella</u> h. <u>tibia</u> i. <u>tarsals</u> j. <u>metatarsals</u> k. <u>coxa</u> l. <u>sacrum</u> m. <u>costal cartilage</u> n. <u>rib</u> o. <u>sternum</u> p. <u>cervical vertebrae</u> q. <u>thoracic vertebrae</u> r. <u>lumbar vertebrae</u> s. <u>pectoral girdle</u> t. <u>pelvic girdle</u>

7. a. <u>frontal bone</u>  b. <u>lacrimal bone</u>  c. <u>zygomatic bone</u>  d. <u>vomer</u>  e. <u>mandible</u> f. <u>temporal bone</u>  g. <u>occipital bone</u>

## SOLUTIONS TO THE TEST FOR MODULE #4

1. a. <u>Osteoblast</u> - A bone-forming cell

   b. <u>Osteocyte</u> - A mature bone cell surrounded by bone matrix

   c. <u>Osteoclast</u> - A large, multinucleated cell that breaks down bone

   d. <u>Hematoma</u> - A localized mass of blood that is confined to an organ or some definable space

   e. <u>Callus</u> - A mass of tissue that connects the ends of a broken bone

   f. <u>Anatomical position</u> - The position acquired when one stands erect with the feet facing forward, the upper limbs hanging at the sides, and the palms facing forward with the thumbs to the outside

2. <u>The bone will be less flexible.</u> You could also say that the bone will be brittle or easy to break.

3. <u>Cancellous bone tissue</u> is made of trabeculae.

4. <u>Canaliculi</u> are extensions of osteocytes which allow the cells in bone tissue to communicate with one another.

5. There are at least 6 reasons. You need only have three.

   a. <u>All new bone tissue is cancellous bone. Some new bone tissue must be compact bone, so cancellous bone tissue often needs to be remodeled to compact bone tissue.</u>

   b. <u>Bones increase and decrease in mass based on the stress they experience.</u>

   c. <u>Bone is remodeled in order to re-shape the bone as needed.</u>

   d. <u>Bone is remodeled to repair broken bones.</u>

   e. <u>Bone is remodeled to replace worn collagen or hydroxyapatite.</u>

   f. <u>Bone is remodeled to regulate the calcium levels in your body.</u>

6. <u>Appositional bone growth</u> can occur when the epiphyseal plate is ossified. If you don't have that term, but you have the fact that the bones can grow thicker, you receive half credit.

7. <u>The external callus</u> helps hold the broken pieces of bone together.

8. The parathyroid gland secretes PTH. Since PTH stimulates osteoclast activity, the person's blood calcium level must be low.

9. The sex hormones increase osteoblast activity, which first stimulates bone growth. This leads to the growth spurt. However, at the same time, they stimulate ossification of the epiphyseal plates, which eventually halts bone growth, at least in terms of length. This causes growth to stop.

10. Fibrous joints and cartilaginous joints offer little or no motion.

11. The ball and socket joint offers the greatest range of motion.

12. The gliding joint (or plane joint) offers the least amount of motion.

13. a. This is flexion.
b. This is dorsiflexion.
c. This is circumduction.

14. a. fibrous capsule    b. synovial membrane    c. synovial fluid    d. articular cartilage

## SOLUTIONS TO THE TEST FOR MODULE #5

1. a. <u>Sarcomere</u> - The repeating unit of a myofibril

b. <u>Motor unit</u> - One motor neuron and all the muscle fibers it innervates

c. <u>All-or-none law of skeletal muscle contraction</u> - An individual muscle fiber contracts with equal force in response to each action potential.

d. <u>Muscle tone</u> - The state of partial contraction in a muscle, even when the muscle is not being used

2. This is <u>skeletal muscle tissue</u>. Cardiac (heart) muscle tissue is also acceptable.

3. a. <u>epimysium</u>   b. <u>muscle cell (fiber)</u>   c. <u>fascicle</u>

4. a. <u>actin myofilament</u>   b. <u>myosin myofilament</u>   c. <u>H zone</u>   d. <u>A band</u>

5. The following remain the same: <u>a, b, and d</u>.

6. <u>acetylcholinesterase</u> and <u>ACh</u>

7. <u>It travels down a T-tubule</u>.

8. <u>The cell is beginning to contract</u>, because calcium ions must bind to the troponin on an actin myofilament to start the contraction process. If calcium ions are DIFFUSING, they are moving OUT of the sarcoplasmic reticulum.

9. Between the power stroke and the return stroke, <u>ATP binds to the myosin heads, causing them to release the active sites</u>.

10. <u>ACh must be released by the presynaptic terminal, and the ACh must travel across the synaptic cleft</u>. You can also include the fact that ACh must interact with the cell membrane.

11. <u>No</u>. If they were a part of the same motor unit, they would contract at the same time.

12. If a muscle is stiff, the myosin heads will not let go of the active sites. This means there is not enough <u>ATP</u> in the muscle cells to bind to the myosin heads so that they will release the active sites.

13. <u>Yes</u>. Submaximal means that some but not all of the motor units have been recruited.

14. <u>They all contract</u>.

15. <u>Anaerobic respiration is occurring</u>. Lactic acid is a byproduct of anaerobic respiration.

16. <u>This does not require oxygen.</u>  It is an alternative to aerobic respiration which does not need to wait for oxygen to be transported to the cell.

## SOLUTIONS TO THE TEST FOR MODULE #6

1. a. <u>Origin</u> - The point at which a muscle's tendon attaches to the more stationary bone

   b. <u>Insertion</u> - The point at which a muscle's tendon attaches to the moveable bone

   c. <u>Belly</u> - The largest part of the muscle, which actually contains the muscle cells

   d. <u>Mastication</u> - The process of chewing

2. a. <u>sternocleidomastoid</u>   b. <u>pectoralis major</u>   c. <u>serratus anterior</u>   d. <u>external oblique</u>
   e. <u>sartorius</u>   f. <u>rectus femoris</u>   g. <u>vastus lateralis</u>   h. <u>vastus medialis</u>   i. <u>frontalis</u>
   j. <u>orbicularis oculi</u>   k. <u>orbicularis oris</u>   l. <u>deltoid</u>   m. <u>biceps brachii</u>   n. <u>brachioradialis</u>
   o. <u>rectus abdominis</u>   p. <u>tibialis anterior</u>   q. <u>temporalis</u>   r. <u>occipitalis</u>   s. <u>trapezius</u>
   t. <u>deltoid</u>   u. <u>triceps brachii</u>   v. <u>biceps femoris</u>   w. <u>gastrocnemius</u>   x. <u>latissimus dorsi</u>
   y. <u>gluteus medius</u>   z. <u>gluteus maximus</u>   aa. <u>Achilles tendon</u> or <u>calcaneal tendon</u>

3. The <u>buccinator</u> is the other kissing muscle.

4. The <u>rectus abdominus, external oblique, and internal oblique</u> muscles work on the vertebral column. The internal oblique is not in the figure, but we would count it. Only one is needed.

5. The forearm flexors are: <u>biceps brachii, brachioradialis, and pronator teres</u>. The antagonist is the extensor: <u>triceps brachii</u>. Once again, go ahead and count the ones that are not on the figure. Only one of each is needed.

6. The <u>Achilles tendon</u> or <u>calcaneal tendon</u> and the <u>extensor retinaculum</u> are not muscles.

7. The thigh flexors are: <u>iliacus, psoas major, rectus femoris, sartorius, and adductor longus</u>. The student needs only list one. The antagonists are the extensors: <u>gluteus maximus, semitendinosus, semimembranosus, biceps femoris, and adductor magnus</u>. The student need list only one, and count any that are not on the figure.

# SOLUTIONS TO THE TEST FOR MODULE #7

1. a. <u>Ganglia</u> - Collections of neuron cell bodies which are outside of the CNS

b. <u>Spinal nerves</u> - Nerves which originate from the spinal cord

c. <u>Efferent neurons</u> - Neurons which transmit action potentials from CNS to the effector organs

d. <u>Somatic motor nervous system</u> - The system that transmits action potentials from the CNS to the skeletal muscles

e. <u>Association neuron</u> - A neuron that conducts action potentials from one neuron to another neuron within the CNS

f. <u>Excitability</u> – The ability to create an action potential in response to a stimulus


2. a. <u>presynaptic terminals</u>   b. <u>node of Ranvier</u>   c. <u>Golgi apparatuses</u>   d. <u>nucleus</u>
   e. <u>dendrites</u>   f. <u>cell body</u>   g. <u>axon hillock</u>   h. <u>axon</u>

3. All neurons have <u>only one axon</u>.

4. Those cells secrete cerebrospinal fluid.

5. The <u>autonomic nervous system</u> controls smooth muscles.

6. In order for a severed axon to regenerate, <u>the axon must be covered with Schwann cells and the axon must be reasonably well-aligned with its severed end</u>. A CNS axon is covered by oligodendrocytes; thus, <u>CNS axons cannot regenerate when severed</u>.

7. During depolarization, <u>the sodium ions are rushing into the cell</u>. This makes the potential difference more positive. <u>Nothing is really happening to the potassium ions</u>. Some potassium ions are leaking out of the cell because the membrane is naturally permeable to potassium ions, but that's all.

8. If the potassium ion concentration is high outside the axon and the sodium ion concentration inside the cell is high, then repolarization has occurred. This means that the <u>potential difference is negative</u>. In fact, it is probably more negative than the resting potential.

9. In order to get back to the resting potential, the <u>sodium-potassium exchange pump must send the sodium outside of the axon and the potassium inside the axon</u>.

10. <u>The stimulus might be subthreshold, or the axon might be in its absolute refractory period</u>.

11. <u>This is called saltatory conduction, and it is faster than continuous conduction</u>.

12. There is no difference between the maximum potential difference of each individual action potential. Action potentials work on the all-or-nothing principle. Thus, each individual action potential looks the same, regardless of the stimulus involved.

13. The frequency of action potentials is greater AFTER your little brother turned up the radio as compared to before. The strength of a signal running along a nerve is determined by the frequency of action potentials. The stronger the stimulus, the higher the frequency of action potentials.

14. This signal did not travel through a synapse. Synapses regulate the signal by either reducing the number of action potentials (excitatory synapse) or inhibiting the signal (inhibitory synapse). If the frequency of action potentials is exactly the same from beginning to end, it must never have passed through a synapse.

15. An excitatory synapse requires several signals from the presynaptic axon in order to generate an action potential on the postsynaptic neuron. Thus, there will be fewer action potentials on the other side of the synapse. This means the frequency of action potentials on the postsynaptic neuron will be LOWER than the frequency of the action potentials on the presynaptic neuron.

16. This is an inhibitory synapse. The opening of potassium channels will make the potential difference even more negative. This will move the neuron away from threshold, so it will make the neuron less likely to produce an action potential.

17. The signal must have gone through a divergent circuit, since one signal produced many signals.

# SOLUTIONS TO THE TEST FOR MODULE #8

1. a. <u>Gray matter</u> – Collections of nerve cell bodies and their associated neuroglia

   b. <u>White matter</u> – Bundles of parallel axons and their sheaths

   c. <u>Decussation</u> – A crossing over

   d. <u>Vital functions</u> – Those functions of the body necessary for life on a short-term basis

   e. <u>Commissures</u> - Connections of nerve fibers which allow the two hemispheres of the brain to communicate with one another

2. a. <u>gyri</u>  b. <u>sulci</u>  c. <u>cerebrum</u>  d. <u>thalamus</u>  e. <u>midbrain</u>  f. <u>pons</u>  g. <u>medulla</u>  h. <u>cerebellum</u>

3. <u>The brainstem is formed by the medulla, the pons, and the midbrain.</u>

4. <u>The diencephalon is formed by the thalamus and the hypothalamus.</u>

5. <u>The ventricles produce cerebrospinal fluid.</u>

6. <u>A reflex arc allows your muscles to react more quickly than they would if the brain were to have to make all the decisions regarding working the muscles.</u>

7. <u>In a reflex arc, an afferent neuron sends a signal to an association neuron in the spinal cord. If the signal is interpreted as severe pain, heat, or something like that, the association neuron immediately sends a signal down the efferent neuron which controls the muscles.</u>

8. <u>The association neuron in the reflex arc forms a converging circuit which synapses at the cell body. That way, both the reflex arc and the brain can control the muscle.</u>

9. <u>i</u>

10. <u>d</u>

11. <u>h</u>

12. <u>e</u>

13. <u>b</u>

14. <u>c</u>

15. l

16. g

17. j

18. k

19. f

20. a

# SOLUTIONS TO THE TEST FOR MODULE #9

1. a. <u>Sensory receptor</u> - An organ which responds to a specific type of stimulus by ultimately triggering an action potential on a sensory neuron

b. <u>Thermoreceptors</u> - Sensory receptors which respond to heat or cold

c. <u>Nociceptors</u> - Sensory receptors which respond to pain or excess stimulation

d. <u>Cutaneous receptors</u> - Receptors in the skin

e. <u>Proprioceptors</u> - Receptors in the muscles and tendons

2. Since there are two neurons, it is a part of the <u>ANS</u>. More specifically, it is in the <u>parasympathetic division</u>, because the first neuron is long.

3. There are several. The student need only list three:

<u>Free nerve endings: receptors for heat, cold, movement, itch, and pain</u>
<u>Merkel's disks: receptors for light touch</u>
<u>Hair follicle receptors: receptors that detect the movement of hair</u>
<u>Pacinian corpuscle: pressure receptors</u>
<u>Meissner's corpuscles: two-point discrimination</u>
<u>Ruffini's organ: pressure and stretch receptors</u>

4. <u>The substance must be volatile because the substance must get into the nasal cavity and up to the olfactory recess. Only gases can do that. The substance must be partly water-soluble to pass through the mucous layer. The substance must be partly lipid soluble to get through the cell membrane.</u>

5. a. <u>auricle</u>   b. <u>external auditory meatus</u>   c. <u>tympanic membrane</u>   d. <u>auditory ossicles</u>
e. <u>semicircular canals</u>   f. <u>cochlea</u>

6. The cupula is involved in the sense of dynamic equilibrium and is housed in the <u>semicircular canals</u>.

7. The organ of hearing is in the <u>cochlea</u>.

8. The vibrations start at the ear drum (tympanic membrane). That vibrates the malleus, then the incus, and then the stapes. They vibrate the oval window, which vibrates the perilymph. The perilymph vibrates the basilar membrane, which vibrates the endolymph, causing the tectorial membrane to vibrate, which vibrates the hair cells. Thus, <u>the ear drum is first, and the hair cells are last</u>. If the student said that the tectorial membrane is last, give him or her credit.

9. a. lens  b. sclera  c. optic nerve  d. vitreous humor  e. retina  f. choroid  g. conjunctiva  h. posterior chamber  i. anterior chamber  j. cornea  k. pupil  l. iris  m. suspensory ligaments  n. ciliary body

10. The lens, ciliary body, and suspensory ligaments are involved in accommodation, since accommodation is the process by which the lens adjusts to change focus based on the distance of the image.

11. The cornea and lens are the two most important structures in the eye for bending light so that it focuses properly.

12. Rods and cones are found in the retina. Cones are concentrated in the fovea centralis.

13. The ultricular macula should be investigated. The crista ampullaris and the ampullae are both related to dynamic equilibrium. The external auditory meatus is involved in hearing, not balance.

# SOLUTIONS TO THE TEST FOR MODULE #10

1. a. <u>Neurosecretory cells</u> - Neurons of the hypothalamus that secrete neurohormone rather than neurotransmitter

   b. <u>Prostaglandins</u> - Biologically active lipids which produce many effects in the body, including smooth muscle contractions, inflammation, and pain

2. a. <u>hypothalamus</u>  b. <u>pituitary</u>  c. <u>thyroid</u>  d. <u>adrenals</u>  e. <u>ovaries in females</u>  f. <u>pineal body</u>  g. <u>parathyroids</u>  h. <u>thymus</u>  i. <u>pancreas</u>  j. <u>testes in males</u>

3. This is <u>constant secretion</u>. The minor fluctuations are due to the body maintaining homeostasis.

4. <u>i, released from the hypothalamus</u>

5. <u>k, released from the parathyroid</u>

6. <u>m, released from the adrenal cortex</u>

7. <u>j, released from the testes</u>

8. <u>e, released from the anterior pituitary</u>

9. <u>g, released from the anterior pituitary</u>

10. <u>f, released from the pineal body</u>

11. <u>l, released from the adrenal medulla</u>

12. <u>a, released from the posterior pituitary</u>

13. <u>b, released from the hypothalamus</u>

14. <u>c, released from the hypothalamus</u>

15. <u>h, released from the posterior pituitary</u>

16. <u>d, released from the anterior pituitary</u>

17. <u>p, released from the pancreas</u>

18. <u>q, released from the ovaries</u>

19. <u>n, released from the adrenal cortex</u>

20. o, released from the pancreas

21. This hormone stimulates a membrane-bound receptor.  No new protein is being made.  Thus, this is not an intracellular receptor.

# SOLUTIONS TO THE TEST FOR MODULE #11

1 a. <u>Erythrocytes</u> - Red blood cells which carry the oxygen in blood

b. <u>Leukocytes</u> - White blood cells which perform various defensive functions in blood

c. <u>Platelets</u> - Cell fragments in blood which help prevent blood loss

d. <u>Hemopoiesis</u> - The process by which the formed elements of blood are made in the body

e. <u>Systolic phase</u> - The phase of the cardiac cycle in which the ventricles contract

f. <u>Diastolic phase</u> - The phase of the cardiac cycle in which the ventricles relax

g. <u>Arterioles</u> - The smallest arteries that still have three tunics

h. <u>Venules</u> - Small veins that do not have three tunics but instead have only an endothelium, a basement membrane, and a few smooth muscle cells

2. <u>b</u>

3. <u>c</u>

4. <u>f</u>

5. <u>d</u>

6. <u>g</u>

7. <u>b</u>

8. <u>a</u>

9. <u>e</u>

10. Since prothrombinase is present, that means blood coagulation is occurring. Thus, the <u>blood coagulation phase</u> is occurring. If the student answered third stage, that's fine as well. Please note that the question DID NOT ask what stage of the blood coagulation process was occurring. It asked what stage of hemostasis is occurring. Hemostasis has three stages: vascular constriction, platelet plug formation, and coagulation.

11. In general, blood coagulation factors are always in the blood. However, the activated factors are only in the blood during coagulation. Calcium ions are always in the blood, they just help the coagulation process once it begins. Thrombin and fibrin are made as a result of the coagulation

process, but fibrinogen is always there. Thus, the chemicals that exist only during blood coagulation are: <u>activated factor XII, thrombin, and fibrin</u>.

12. Type AB blood has both A and B antigens and thus cannot form antibodies against these antigens. If the blood is also Rh-positive, then it has the Rh antigen and cannot make antibodies against it. Thus, <u>AB+</u> blood has no antibodies against erythrocyte antigens or Rh antigens. Note that most Rh- people do not form anti-Rh antibodies unless exposed to the Rh antigen.

13. If the blood has antibodies against the B antigen, it cannot be B or AB blood. O has antibodies against the B antigen, but it also has antibodies against the A antigen. Thus, the type is A. If it has antibodies against the Rh antigen, then it cannot be Rh positive. Thus, this blood is <u>A-</u>. Note that most Rh- people do not form anti-Rh antibodies unless exposed to the Rh antigen.

14. If the father is AB, then he has an A allele and a B allele. If the mother us type A, she is either AA or AO. Since the child is B, the child must have gotten the B allele from the father. The only way to be type B, then, is to get an O from the mother. The mother, then, is AO. Since Rh-negative is recessive, the child must have two Rh-negative alleles. Thus, the child must have inherited one from each parent. The only way the father can have a Rh-negative allele and be Rh positive is to be heterozygous. Thus, <u>the father has AB alleles and +- alleles. The mother has AO and -- alleles</u>.

15. a. <u>superior vena cava</u>   b. <u>pulmonary semilunar valve</u>   c. <u>right atrium</u>
d. <u>right atrioventricular canal</u>   e. <u>right atrioventricular valve</u> or tricuspid valve
f. <u>chordae tendineae</u>   g. <u>right ventricle</u>   h. <u>aorta</u>   i. <u>left atrium</u>   j. <u>aortic semilunar valve</u>
k. <u>left atrioventricular canal</u>   l. <u>left atrioventricular valve</u> or bicuspid valve   m. <u>left ventricle</u>

16. <u>The atria are relaxed, the ventricles are contracted, both atrioventricular valves are closed, and both semilunar valves are open.</u>

17. <u>oxygenated</u>
<u>deoxygenated</u>
<u>right atrium</u>
<u>right atrioventricular canal</u> (can also come AFTER the one below)
<u>right atrioventricular valve</u>
<u>right ventricle</u>
<u>pulmonary semilunar valve</u>
<u>oxygenated</u>
<u>left atrium</u>
<u>left atrioventricular canal</u> (can also come AFTER the one below)
<u>left atrioventricular valve</u>
<u>left ventricle</u>
<u>aortic semilunar valve</u>

# SOLUTIONS TO THE TEST FOR MODULE #12

1. a. <u>Lymph nodes</u> – Encapsulated masses of lymph tissue found along lymph vessels

b. <u>Diffuse lymphatic tissue</u> – Concentrations of lymphatic tissue with no clear boundaries

c. <u>Innate immunity</u> - An immune response that is the same regardless of the pathogen or toxin encountered

d. <u>Acquired immunity</u> - An immune response targeted at a specific pathogen or toxin

e. <u>Humoral immunity</u> - Immunity which comes from antibodies in blood plasma

f. <u>Cell-mediated immunity</u> - Immunity which comes from the actions of T-lymphocytes

2. k

3. d

4. o

5. h

6. a

7. s

8. n

9. r

10. b

11. c

12. q

13. e

14. p

15. f

16. m

17. l

18. g

19. j

20. i

# SOLTUIONS TO THE TEST FOR MODULE #13

1. a. <u>Deglutition</u> - The act of swallowing

b. <u>Peristalsis</u> - The process of contraction and relaxation of smooth muscles which pushes food through the alimentary canal

d. <u>Lumen</u> - The hole in the center of a tube

c. <u>Micronutrients</u> – The nutrients the body needs in small amounts, such as vitamins and minerals

2. a. <u>parotid salivary gland</u>  b. <u>esophagus</u>  c. <u>liver</u>  d. <u>transverse colon</u>  e. <u>ascending colon</u>  f. <u>descending colon</u>  g. <u>rectum</u>  h. <u>large intestine</u>  i. <u>sublingual salivary gland</u>  j. <u>submandibular salivary gland</u>  k. <u>stomach</u>  l. <u>pancreas</u>  m. <u>gall bladder</u>

3. a. <u>liver</u>  b. <u>salivary glands</u>  c. <u>pancreas</u>  d. <u>liver</u>  e. <u>stomach</u>  f. <u>rectum</u>  g. <u>small intestine</u>  h. <u>pancreas</u>  i. <u>small intestine</u>  j. <u>small intestine</u>  k. <u>gall bladder</u>  l. <u>esophagus</u>  m. <u>liver</u>

4. <u>The epiglottis closes off the larynx to prevent food from going down the wrong pipe, and the soft palate seals the nasal cavity to prevent breathing.</u>

5. <u>Intestinal villi are tiny projections in the small intestine that radically increase its surface area.</u> The student could also say that they are the structures through which nutrients are absorbed.

6. <u>Cells cannot produce an enzyme that breaks down proteins. It would kill the cell.</u> The cells in the stomach produce pepsinogen, which is activated into pepsin when it enters the gastric juice.

7. <u>The body cannot absorb vitamin $B_{12}$ without intrinsic factor.</u>

8. The macronutrients are <u>carbohydrates, fats, and proteins</u>. The student can say "lipids" instead of fats.

9. <u>Vitamins D and K can be produced in the body.</u>

# SOLUTIONS TO THE TEST FOR MODULE #14

1. a. <u>Lower respiratory tract</u> – The part of the respiratory system containing the larynx, trachea, bronchi, and lungs

b. <u>Ventilation</u> – The process of getting air into the lungs and getting it back out

c. <u>Internal respiration</u> – The process of $O_2$ and $CO_2$ exchange between the cells and the blood

d. <u>Pneumothorax</u> - Air in the pleural cavity, which leads to a collapsed lung

e. <u>Compliance</u> - The ease with which the lungs inflate

f. <u>Tidal volume</u> - The volume of air inhaled or exhaled during normal, quiet breathing

g. <u>Functional residual capacity</u> - The volume of air left in the lungs after a normal exhalation

2. a. <u>nasal cavity</u>   b. <u>pharynx</u>   c. <u>larynx</u>   d. <u>trachea</u>   e. <u>bronchi</u>   f. <u>diaphragm</u>

3. <u>c</u>

4. <u>d</u>

5. <u>a</u>

6. <u>b</u>

7. If the diaphragm is contracted and air is flowing, the person is inhaling. Thus, the volume in the thoracic cavity is large and thus the pressure in the lungs is low. Therefore, <u>the pressure in the lungs is lower than that of the atmosphere</u>.

8. The abdominal muscles are muscles of forced expiration. Thus, the person is <u>forcefully exhaling</u>.

9. The sternocleidomastoid and scalene muscles are the muscles of forced inspiration. Thus, this person is inhaling as much as possible. The lungs, therefore, will fill to the <u>total lung capacity</u>.

10. If the airway is blocked, not enough air gets to the lungs. However, the blood does not slow down. Thus, <u>the controlled relationship between ventilation and blood flow</u> is affected.

11. The <u>basement membrane of the alveolar epithelium and the basement membrane of the capillary endothelium</u> were left out.

12. <u>They send signals to the medulla</u>, where the respiratory control centers are located. The signals are <u>inhibitory</u>, since they prevent over-inflation of the lungs.

13. The respiratory system can <u>increase the rate and depth of ventilation</u>. This will decrease the amount of $CO_2$ in the blood, which will decrease the amount of carbonic acid in the blood. That will, in turn, increase the blood pH.

14. a. <u>glycolysis</u>  b. <u>pyruvate</u>  c. <u>ATP</u>  d. <u>4</u>  e. <u>NADH</u>  f. <u>oxidation of pyruvate</u>
g. <u>NAD+</u>  h. <u>Kreb's cycle</u> (Citric acid cycle is fine as well.)  i. <u>oxaloacetic acid</u>  j. <u>$CO_2$</u>
k. <u>$FADH_2$</u>  l. <u>electron transport system</u>  m. <u>NADH</u>  n. <u>$FADH_2$</u> (m and n could be switched)
o. <u>$H^+$ or hydrogen ions</u>   p. <u>ATP</u>

# SOLUTIONS TO THE TEST FOR MODULE #15

1. a. <u>Retroperitoneal</u> – Behind the parietal peritoneum

b. <u>Erythropoiesis</u> – The production of red blood cells (erythrocytes)

c. <u>Renal blood flow rate</u> - The rate at which blood flows through the kidneys (1 liter/min)

d. <u>Glomerular filtration rate</u> - The rate at which filtrate is produced in glomerular filtration (125 mL/minute)

e. <u>Tubular maximum</u> - The maximum rate of reabsorption by active transport through the nephron tubules

f. <u>Buffer system</u> - A mixture of an acid and a base which resists changes in pH

2. The filtrate runs into these structures in this order:

<u>proximal tubule, loop of Henle, distal tubule, collecting duct, minor calyx, major calyx, and renal pelvis</u>

3. a. <u>renal corpuscle</u>   b. <u>Bowman's capsule</u>   c. <u>glomerulus</u>   d. <u>descending limb of the loop of Henle</u>   e. <u>loop of Henle</u>   f. <u>ascending limb of the loop of Henle</u>
g. <u>proximal tubule</u>   h. <u>distal tubule</u>   i. <u>collecting duct</u>

4. <u>The distal tubule and collecting duct change their permeability to water. The change is determined by the amount of ADH present.</u>

5. a. The arrow shows solutes traveling from the nephron to the blood. This is <u>reabsorption</u>.

b. The arrow shows solutes traveling from the blood to the nephron. This is <u>secretion</u>.

6. GCP is glomerular capillary pressure. Increasing that will <u>increase GFR</u>, which is the glomerular filtration rate.

7. Since substance A has a higher T-max, it will be reabsorbed more in the blood. Thus, <u>the concentration of A is greater in the blood than the concentration of B.</u>

8. <u>Juxtaglomerular cells are cells in the kidney which sense and respond to low blood pressure and low sodium concentration in the blood. They secrete renin, which stimulates a sequence of events that increase blood pressure and increases sodium concentration in the blood.</u>

9. The proper range of blood pH is <u>7.35 to 7.45</u>.

10. The pH is below normal, so this is <u>acidosis</u>.

11. The body has three systems for controlling blood pH. <u>The body has buffer systems which resist the change in pH. They are very fast but less effective than the other two. The body can also control blood pH by changing the depth and rate of ventilation. This is more effective than the buffer systems, but it is a bit slower. Finally, the body controls blood pH by secreting $H^+$ in the nephron. This is the most effective but slowest process.</u>

12. An acid will cause the base in the buffer to react. The base in the bicarbonate buffer is <u>bicarbonate</u>. The student could also give the chemical formula, $HCO_3^-$.

# SOLUTIONS TO THE TEST FOR MODULE #16

1. a. <u>Spermatogenesis</u> – The process by which sperm form in the testes

b. <u>Semen</u> – A milky-white mixture of sperm and the secretions of the testes, seminal vesicles, prostate gland, and bulbourethral glands

c. <u>Secondary sex characteristics</u> – The characteristics which appear at puberty and tend to distinguish men from women. These include the development of breasts, hairline patterns, facial shape, body shape, and the distribution of body hair.

d. <u>Puberty</u> – A series of events which transforms a child into a sexually mature adult

e. <u>Anabolism</u> – All of the synthesis reactions which occur in the body

f. <u>Catabolism</u> – All of the decomposition reactions which occur in the body


2. a. <u>vas deferens</u>   b. <u>erectile tissue</u>   c. <u>penis</u>   d. <u>glans penis</u>   e. <u>foreskin or prepuce</u>
f. <u>seminal vesicle</u>   g. <u>prostate gland</u>   h. <u>bulbourethral gland</u>   i. <u>urethra</u>   j. <u>epididymis</u>
k. <u>testis</u>   l. <u>ovary</u>   m. <u>uterus</u>   n. <u>cervix</u>   o. <u>clitoris</u>   p. <u>vagina</u>

3. <u>e</u>

4. <u>f</u>

5. <u>a</u>

6. <u>c</u>

7. <u>d</u>

8. <u>g</u>

9. <u>b</u>

10. <u>They each contribute equal amounts.</u>

11. <u>No, they do not contribute equal amounts. The egg cell contributes virtually all of the cytoplasm.</u>

12. <u>Ovulation has occurred.</u> FSH and LH decrease during the luteal stage.

13. <u>Progesterone</u> dominates the luteal phase.

14. The proper order is: <u>zygote, cleavage, morula, blastula, gastrula, neurula</u>

15. The <u>zygote, cleavage, and morula</u> stages are completed outside of the uterus. Implantation occurs during the blastula stage, however, so it cannot be completed outside of the uterus.